SEX AND FRIENDSHIP IN BABOONS

SEX AND FRIENDSHIP IN BABOONS

With a New Preface

BARBARA B. SMUTS

HARVARD UNIVERSITY PRESS
Cambridge, Massachusetts
London, England

Printed in the United States of America

First Harvard University Press paperback edition, 1999
Originally published by Aldine de Gruyter, a division of Walter de Gruyter, Inc., 200 Saw Mill Road, Hawthorne, N.Y. 10532

Library of Congress Cataloging-in-Publication Data
Smuts, Barbara B.
Sex and friendship in baboons / Barbara B. Smuts.
p. cm.
Originally published: New York : Aldine Pub. Co., c1985.
Includes bibliographical references (p.).
ISBN 0-674-80275-6
1. Baboons—Behavior. 2. Sexual behavior in animals.
3. Social behavior in animals. I. Title.
QL737.P93S58 1999
599.8′65156—dc21 99-43248

In Memory of Alex and Daphne

PREFACE, 1999

When we say that we have a personal relationship with someone, we mean that we have a bond based on our individual identities, a relationship in which who I am and who you are makes a difference to how we are with each other. This book demonstrates that baboons also form such personal relationships. Adult females pair up with adult males in distinctive partnerships that stand out from the vast majority of female-male relationships. Sometimes these partnerships are intense but brief. Sometimes they last for much of an adult lifetime, sliding into an easy intimacy characterized by the freedom to ignore each other. Sometimes two adolescents, both new to the complicated world of adult sexuality and parenthood, become partners. Sometimes an old female, wise in the ways of males, bonds with a fierce, newly mature male who follows her around with puppy-like devotion. Sometimes two aging partners feed and rest in companionable silence, as they have done day in and day out for many years.

Some friendships are calm and low-key, leaving plenty of opportunity for each partner to socialize with other members of the opposite sex. Others are volatile, full of jealousy and periods of cold-shouldering, followed by reconciliations and good times. Some pairs spend much time together but rarely touch, while others groom frequently, embrace when their paths cross, and huddle together at night.

These relationships change as each baboon goes through different phases of life and matures into an ever-more idiosyncratic individual. Some relationships dissolve: the male leaves the troop for greener pastures; the female falls for someone else during a series of intense sexual encounters; the male gains in status and drops his adolescent friend for a new bond with the alpha female. Others settle into comfortable, relaxed familiarity.[1]

The reasons baboons form friendships with members of the opposite sex seem almost as diverse as the many forms these friendships take.[2] Young adult males who have just entered a troop in which strength doesn't guarantee sex cultivate bonds with females, both as a way to penetrate a complex network of interlocking ties and as a way to entice females into mating with them later.[3] Older males that have already consorted with several females maintain bonds with some of them through the many months of pregnancy and lactation, long after the females have lost their sexual appeal.[4] These males befriend the females' tiny infants, protecting them from predators and from other baboons, perhaps especially against other males contemplating infanticide.[5] Maybe the male is protecting his genetic stake.[6] Maybe he is showing the female that if she consorts with him again the next time, he will stick around and help her. Maybe he is doing both.[7]

A female may go for a male because he invites her and her infant to forage near him, keeping other males and higher-ranking females away without asking much in return. Another female may prefer a recent immigrant because his ability to intimidate every other male in the group makes her feel relatively safe around him, even though he sometimes beats her up himself. Still another favors an especially calm male who grooms her infant and plays with it gently while she feeds in another tree. And another female, an adolescent, may hang out with a particular male because he is friends with her sister and her cousin, and she wants to keep on spending time with them.

Researchers will probably continue to debate for many years the reasons (in terms of evolutionary advantages) that baboon males and females form friendships. But more important, I think, than the exact reproductive advantages resulting from friendships is that they exist in every baboon troop that has ever been studied carefully and that they are so personal and so variable. The fact that this is so is not a coincidence, for it is the personal nature of the relationships that leads each one to be so distinctive, just as human marriages differ radically from one another. Discovering that baboons have friendships is just the beginning. This discovery opens the door to a vast series of questions about why this baboon ends up with that one, and another one with someone else, and about how and why the quality of each relationship is so different. These questions will yield answers that may ultimately be amenable to evolutionary analysis, but before that occurs, researchers must delve deeply into the richness of baboon lives and the nuances of their behavior. What do we know now, that we did not know in 1985, about how baboons relate to one another and why they do what they do?

We know that different males bring very different tools to the male-male psychological warfare I describe in this book. Some males have a knack for knowing when to pick fights they can win and for taking their frustrations out on someone else when they occasionally lose. These males tend to keep cool, internally, until and unless the situation demands a rapid response, and then they're ready for action ("fighters").

Other males concentrate on forming sexual consortships and are especially attentive to their sexual partners ("lovers"), while still others put a lot of energy into cultivating bonds with mothers and infants ("friends"). The males who successfully pursue any of these strategies experience lower stress than do males with less highly developed social skills.[8]

We also know that when a male olive baboon transfers to a new troop for the first time, he faces a complex series of social challenges that different males meet in very different ways. Some males concentrate on dominating resident males, whereas others tend to avoid aggression and focus instead on making friends with females. Some find a home in the first troop they enter, but others fail and try again in another troop.[9] After spending several years in a troop, each male must decide whether to remain for good or transfer once again. If he has lots of friendly interactions with females and infants, he's considerably more likely to stay.[10]

We know that when a male forms a friendship with a female, he becomes highly attuned to her screams for help, especially if she has a young infant and especially if he hears the aggressive voice of a recently immigrated male at the same time. The screams of females who are not his friends are much less likely to stir his attention.[11]

We know that in some troops, at least some of the time, the highest-ranking male somehow manages to monopolize sexual access to most females right around the time they conceive, no matter with whom those females might want to mate.[12] Yet in other troops, many males, including lower-ranking males, consort with fertile females,[13] and it seems that females manipulate consort opportunities and often end up mating with their preferred partners.[14] In these contexts, different females choose different males as friends; they also differ as to which males they prefer as lovers.[15] This suggests that female preferences are not primarily based on attributes we can easily measure—like rank, age, or length of residence in the troop—but rather on specific characteristics of particular males and on the history of their relationships with those males. But we still haven't a clue why a female prefers the male(s) that she does!

Finally, we also know much more about male-female affiliation in other primates than we did in 1985. Special male-female relationships reminiscent of baboon friendships occur among other primates, including rhesus and Japanese macaques, ringtailed and red-fronted lemurs, and perhaps even chimpanzees.[16] It also appears that in many primates, especially high rates of mutual, friendly interactions occur among particular female-male pairs, even though friendships may not necessarily occur with the regularity we see in baboons.[17] We know that both male sexual coercion and infanticide are widespread in many primates, and in virtually all of these species, affiliated males protect females and/or their offspring from conspecifics, including infanticidal males.[18] And just as in baboons, males appear sometimes to invest in these relationships in order to enhance mating success, while at other times, they seem likely to be investing in their own infants.[19]

These recent findings provide some tantalizing glimpses into sex and friendship in baboons and other species, but we still know surprisingly little about differences between one friendship and another or how friendships develop and change over time, or how friends communicate their feelings and attitudes toward one another. Such matters remain mysterious because most research on male-female relations continues to focus on functional questions (such as whether males are the fathers of the infants they befriend) and overall patterns (such as whether friends spend more time together after the female gives birth), rather than on more proximate questions (how do friendships develop?) or variations within overall patterns (why do some males react more strongly to playbacks of their friends' screams than do others?).

Why is this so? Is it because most scientists find the functional questions and general patterns more interesting or more important? Is it because studying the subtle dynamics of social relationships is often more difficult and more time-consuming? Do young scientists feel that they are more likely to succeed in a competitive market if they focus on functional questions? Is it due to decisions about what gets funded, or what gets published? I don't know the answers to these questions, but I am convinced that scientists need to spend more time understanding how their subjects see the world, how and why their personalities differ, and what makes for an especially close or valuable relationship in a particular society.[20]

In the years since I wrote this book, this conviction has grown. I did not observe baboons again for any length of time until 1993, when David Gubernick and I spent five months studying and videotaping olive baboons at Gombe National Park, Tanzania. This occurred after an interval studying captive chimpanzees[21] and wild bottlenose dolphins[22]—perhaps the two most intelligent and socially complex nonhuman animals on the planet. Yet when I returned to baboons, their social sophistication did not pale in comparison with these brainy paragons; if anything, I saw more complexity among the baboons as a result of an eye better trained to look for it.

One day, for example, two prime adult males, Apié and London, formed a coalition against another prime male, Randani. Randani was one of those males Robert Sapolsky describes as highly skilled in developing relationships with females and young—what I've labeled a "friend-type" male.[23] He was also chronically uninterested in getting into fights with other males. So Randani ignored the other males' threats, and they continued to move closer, pant-grunting and flashing their canines at him while they stood close together, a seemingly invincible duo. Although Apié and London were trying to provoke Randani into a reaction, neither of them wanted to fight with him, so they frequently jockeyed for position, each trying to remain further away from Randani than his partner. Finally, as they got very close, Randani glanced nonchalantly at them and got up to walk away (consistently his favorite tactic whenever challenged by another male). His path of departure took him ever so slightly closer to Apié than to London, and just as Randani was near, London, standing to the side and slightly behind Apié, reached out and gave Apié a strong push, shoving his partner right into Randani's path. Randani, whose position prevented him from witnessing this act, must have assumed that Apié was in the process of attacking him and so he attacked Apié. They fought briefly and then ran off down the beach. London, meanwhile, calmly sat down and watched the tension erupt between two males who would not be fighting each other but for his perfectly timed shove, a Machiavellian action worthy of a chimpanzee.[24]

My time at Gombe also sensitized me to the fragility of most friendships and made me marvel all the more at those few that survive social and demographic turmoil. The troop we observed at Gombe, "A" troop, had split into three subgroups shortly before we began our research. Many male-female friendships ceased as a result, since the partners ended up in different subgroups. Other friendships remained intact initially but faltered as one new male after another moved into the branch of "A" troop that we were studying. Some of these new males threatened old friendships directly by cultivating bonds with females. One male, Chongo (a "fighter-type" according to Sapolsky's typology), persistently and ruthlessly attacked females whenever a male friend of the female was nearby, forcing the male to either fight or fail to protect his friend. (This strategy worked; Chongo was a superior fighter, and through this method he engaged every male, including the highly reluctant alpha male, in fights that Chongo won. As a result, he became the new alpha male within three months of entering the troop.) As I watched most friendships dissolve, or at least weaken, one stood out: the bond between the oldest male in the group, Bofu, and a late-middle-aged female, Santa Fe. In light of typical female behavior, Santa Fe should have given up her old friend, because she was in estrus and extremely attractive to the younger, stronger males. However, although she consorted mostly with them, she also consorted often with Bofu (who very rarely paired up with any other females), and when she wasn't with another male, she hung out with him. Bofu, for his part, kept effectively coming to her defense and that of her offspring, even though his canines were worn. One day I stumbled upon them in a shady glade, sound asleep and curled up together in "spoon" position. Say what you will about scientific objectivity; I'm certain these two baboons loved each other.

Randani and another male, Sherlock, also taught me much about the persistence with which a baboon will attempt to live out his or her own unique way of being in the world. Although I wrote about baboon personalities in this book, it was Sapolsky who really showed, in considerable detail, how adult male social behavior clusters according to different personality types, regardless of age. Because of my interest in friendships, I was especially intrigued with Sapolsky's description of the "friend-type." I thought about Sherlock, who transferred into Eburru Cliffs from the neighboring Pumphouse troop shortly before my study began. Shirley Strum's description of Sherlock's integration into Eburru Cliffs resembles Sapolsky's description of friend-type males.[25] Sherlock avoided interactions with other males but was very popular with females from the start. During my study, he had more female friends by far (six!) than any other male, even though he was still a relative newcomer. Six years later, Sherlock was still there, still had multiple friends, and was still actively consorting. At that point, he was at least thirteen years old (according to Strum's lower age estimate when he was an adolescent), which is definitely over the hill for a male baboon. But what really amazed me was my brief glimpse of Sherlock, seven years later, when I spent three days with Eburru Cliffs during a short visit to Kenya.[26] Although he looked much older, his distinctive face and tail were unmistakable. I was surprised to see him at the center of a cluster of females, and I was truly astonished when, the next day, I saw him contest a hot consortship. Sherlock, now twenty or older, was beyond the age at which most male baboons die, and well past the age at which they typically remove themselves from social center stage.

When I first met Randani in 1993 at Gombe, I kept thinking of Sherlock, but because they looked a lot alike, I attributed this to their physical resemblance. Then one day I was sitting with a small group of "A" troop females, who were resting beside a stream. Most of the troop were downstream, around a bend that made it impossible to see them. The only other baboons they, or I, could see were a few stragglers upstream who were moving toward us. I was looking at the females when, suddenly, all of their faces lit up. One by one, they began to make the "come hither" face (illustrated on p. 6). I turned to look upstream and saw Randani, sitting about thirty yards away, tucking in his chin and shaking his head and shoulders in a way that most males do only when they are particularly excited (I later realized that all of Randani's come hithers are of this intense type, and it makes me wonder whether this is true for friend-type males in general). Randani kept gesturing at these females for so long and with such enthusiasm that I was tempted to make a face back at him myself. In that moment, I was absolutely certain that he was a friend-type male, and later observations confirmed this strong intuition. Although he was not particularly high-ranking, especially for a prime-aged male and recent immigrant, Randani succeeded in forming a close friendship with the alpha female, Hans, during her pregnancy (a time when Hans was demonstrably wary of all the other males in the troop except her friend, Bofu). When Hans's infant son was born, she made great efforts to remain close to Randani. Within a few days of the infant's birth, she began to do something I had never witnessed before. She scorned Bofu and instead brought her infant to Randani, stuck around for a few moments until the infant, predictably, responded positively to Randani's invitations to climb on him, and then trotted off to feed unencumbered, sometimes out of sight many meters away. Given the dangers of infanticide by recent immigrants at Gombe and the fact that there were half a dozen new males in "A" troop that ranked as high as or higher than Randani, this was an extraordinary act of trust.[27] It seemed to be justified. Randani never let the infant stray more than a foot or two away from him, and he kept all other males away. He even charged the new alpha male, the deeply feared Chongo, when he passed close to Hans's infant.[28]

My appreciation for the vicissitudes of baboon relationships has been greatly enhanced by my studies of their greeting behavior. Greetings between pairs are by far the most common kind of social interaction among baboons, with some dyads greeting as often as twenty times in a single day. All baboons greet, including infants only a few weeks old. Greetings follow a standard pattern based on stereotyped elements derived from mother-infant interactions and mating behavior.[29] Two individuals come together, often exchanging grunts and come-hither expressions. One will usually present his or her hindquarters to the other, and the other will respond by touching the hindquarters with the hands or mouth, or even by mounting. When juveniles or females greet, they often hug. Embraces and other kinds of tender touching also occur between males and females and even, rarely, between adult males.

The anthropologist John Watanabe and I conducted a study of the Eburru Cliffs troop in 1983 that documented striking differences among the greetings of different pairs of males.[30] Our most interesting finding concerned the greetings of Alex and Boz, two central characters in this book. These two constituted the tightest and longest-standing male-male alliance in the troop. They never competed, but instead took turns helping each other take estrous females away from younger, stronger males. This egalitarian partnership was mirrored precisely in their greetings, in which they took great pains to alternate in the subordinate, female-like role of presenting and the dominant, male-like role of responding.

Although our research on male-male greetings was rewarding, it was also frustrating because I sensed that we were missing important behaviors that occurred too quickly for us to record accurately in real time. I was therefore very excited by the opportunity to document baboon behavior on videotape at Gombe. We captured 100 hours of "A" troop's interactions on tape, with a focus on greetings among all age-sex classes.

After watching hundreds of these videotaped interactions, I've discovered that, indeed, when interactions are seen only once, in real time, we miss much. I've found, for example, that my conclusions about which baboon started a fight often change when I watch the interaction in slow motion. I've discovered that baboons touch each other (especially brief, light touches on the torso) and also exchange fleeting facial expressions much more often than I could document in real time. My strong subjective impression that greetings vary in "tone" is being confirmed by our ability to quantify such subtleties as tail posture, mutual eye contact, synchrony of movement, frequency of intimate touching, and the overall pace (slow and smooth versus fast and jerky) of gestural exchanges.

Preliminary analyses suggest that greetings between males and females who are friends tend to involve more eye contact and slower pacing than greetings between non-friends. But of even greater interest to me is the possibility of using these tapes to explore how one friendship differs from another. We have found, for instance, that when Santa Fe and Bofu (the devoted couple described above) come together, they often dispense with the formal hindquarter presentation used by other friend pairs and instead greet face to face. It is my hope that detailed analyses of male-female greetings, as well as greetings involving other age-sex classes, will do much to elucidate how baboons communicate their feelings, attitudes, and perhaps even future intentions in their relationships with one another.

What is the purpose of delving so deeply into the details of social relationships in other species? One reason is scholarly, the sort of rationale scientists put in grant proposals to try to convince funding agencies that their work holds relevance for humans. Indeed, I believe that a rich understanding of social relationships in other species (especially primates and other large-brained, long-lived, highly social species like dolphins, elephants, lions, mongooses, and other social carnivores) is very important to understanding ourselves. We've spent more than ninety-nine percent of our history as a species living in small, face-to-face groups like those of wild primates, which means that, just as for baboons today, virtually all of our social relationships have involved a few dozen individuals. This, in turn, implies that human evolutionary social psychology largely reflects the sorts of situations described here for baboons: making, keeping, and losing friends whose aid is critical to survival and reproductive success.[31] Still in its early days, evolutionary psychology has tended to focus on dramatic adaptive problems, such as finding and competing for mates, often by investigating short-term relationships among college students largely unfamiliar to one another and removed from their families and communities. More recently, evolutionary psychology has begun to focus increasingly on the sorts of questions addressed in this book and in other nonhuman research: relationships in community and group settings, long-term bonds, and the motivations and emotions that underlie these long-term commitments.[32] As this happens, evolutionary psychologists will find studies of other animals increasingly relevant since they are often the source of important new discoveries that also apply to humans.

Although these contributions to human understanding are, in my view, very significant, I believe there is an even more important reason for conducting detailed research on animal relationships and animal personalities. I began this introduction by noting that baboons form *personal* relationships just as we do and that what makes these relationships personal is the fact that individual identities matter. I have recently discussed in some detail what this means to us as a species that holds the fate of other animals in our hands.[33] If baboons (and dolphins and lions and many other species) are capable of relating to one another as individuals, this implies that they are also capable of relating to us that way; my experience and that of many other people proves this is so. This means that our world is replete with nonhuman beings with whom each of us could potentially form personal relationships, each with a unique flavor stemming not just from the characteristics of the two species we represent but also from the unique attributes of each individual.[34] Although rare people exist who are devoted to the welfare of other species in principle, for most people, a sense of caring and responsibility for other species depends on feeling directly connected to them. Jane Goodall and Dian Fossey proved that research that makes other animals come alive as individuals, with whom we could in principle have personal relationships, contributes immensely to this kind of awareness. We scientists are privy to a rare and precious opportunity when we come to know intimately nonhuman animals living in their own worlds. We have a responsibility to those animals to show other people who they *really* are—sentient beings who matter to one another, living lives as full of drama and emotion and poetry as our own. To perceive the planet as populated with billions of such creatures staggers the imagination, but it is true, and if we want the world of the future to retain this richness, we need to become ever more conscious of this reality before it is too late.

NOTES

1. These descriptions are based on olive baboons I observed at Gilgil, Kenya, from 1976 to 1979 and in 1983 and on those I observed at Gombe National Park, Tanzania, in 1993.
2. "Reasons" refers both to evolutionary explanations (in terms of reproductive benefits) and to more immediate motivations of the baboons themselves.
3. Shirley C. Strum, "Agonistic Dominance in Male Baboons: An Alternative View," *International Journal of Primatology,* 3 (1982): 175–202, *Almost Human* (New York: Random House, 1987), and "Reconciling Aggression and Social Manipulation as Means of Competition. 1. Life-History Perspective," *International Journal of Primatology,* 15 (1994): 739–765; and Debbie L. Manzolillo, "Intertroop Transfer by Adult Male *Papio anubis*" (Ph.D. diss., University of California, Los Angeles, 1982).
4. During pregnancy and lactation, female baboons cease to cycle and males show no interest in mating with them until they resume, which in the Eburru Cliffs troop occurs, on average, about fourteen months after birth if the infant survives. See Barbara B. Smuts and Nancy A. Nicolson, "Reproduction in Wild Female Olive Baboons," *American Journal of Primatology,* 19 (1989): 229–246. Toward the middle of each estrous cycle, females form sexual consortships in which they pair up with particular males for anywhere from a few hours to several days. During consortships, females mate only with their consort partners. See Glenn Hausfater, "Dominance and Reproduction in Baboons," *Contributions to Primatology,* 7 (1975).
5. Ryne A. Palombit, Robert M. Seyfarth, and Dorothy L. Cheney, "The Adaptive Value of 'Friendships' to Female Baboons: Experimental and Observational Evidence," *Animal Behaviour,* 54 (1997): 599–614.
6. Jeanne Altmann, *Baboon Mothers and Infants* (Cambridge, Mass.: Harvard University Press, 1980); David M. Stein, "Ontogeny of Infant-Adult Male Relationships during the First Year of Life for Yellow Baboons *(Papio cynocephalus),*" in David M. Taub, ed., *Primate Paternalism* (New York: Van Nostrand Reinhold, 1984), pp. 213–243; Michael E. Pereira, "Agonistic Interac-

tions of Juvenile Savanna Baboons. II. Agonistic Support and Rank Acquisition," *Ethology,* 80 (1987): 152–171; Ronald Nöe and Albertha A. Sluijter, "Reproductive Tactics of Male Savanna Baboons," *Behaviour,* 113 (1990): 117–170; and Fred B. Berkovitch, "Mate Selection, Consortship Formation, and Reproductive Tactics in Adult Female Savanna Baboons," *Primates,* 32 (1991): 437–452.

7. In this book I proposed that males gain two main benefits from friendships with females: enhanced chances of mating with the mother in the future ("mating effort" hypothesis), and opportunities to contribute to the survival of their own offspring ("paternal investment" hypothesis). I suggested that the relative importance of these two benefits varied with male life history: new immigrants or recently mature natal males tend to form friendships as a way to enhance mating opportunities, but as a male's tenure in a troop increases, his friendships tend to reflect previous mating activity so that he ends up affiliating with infants he could have sired (pp. 164–169 and 198–200). In some cases, male-infant affiliation may simultaneously enhance the survival of a male's own infant *and* increase his chances of mating with the mother the next time she conceives. Thus, both hypotheses may apply, sometimes to different males, sometimes to the same male with different females or at different points in time.

   Several subsequent studies, however, have treated these hypotheses as mutually exclusive by arguing that male friendships with females always, and only, involve paternal investment. Berkovitch ("Mate Selection") rejected the mating-effort hypothesis for the neighboring Pumphouse troop because males found in proximity to infants were subsequently equally likely to consort, or to fail to consort, with the infants' mothers. His conclusion is based on the assumption that if friendship enhances consort activity, then males will have a greater than 50 percent chance of consorting with their female friends. This assumption is not valid because there is enormous variation among male baboons in the frequency with which they consort, and many factors other than friendship contribute to this variation. For this reason, the mating-effort hypothesis does *not* predict that males will have a greater than 50 percent chance of consorting with friends (as Berkovitch assumes), nor that females will consort more often with friends than non-friends (as Manzolillo assumes in "Intertroop Transfer"). Rather, as I emphasized (p. 168), it predicts that being friends with a female makes a male more likely to consort with her *than he would be otherwise* (e.g., in my study friendship on average roughly doubled a male's chances of consorting). To evaluate *this* prediction one must compare each male's consorting activity with friends with what is expected given his consorting activity overall. This method controls for the variation across males in how frequently they consort. None of the studies claiming to reject the mating-effort hypothesis frames the hypothesis in these terms and none provides a rationale for framing it in another way. Thus, unfortunately, this hypothesis remains untested for other troops or populations.

   Berkovitch ("Mate Selection"), Pereira ("Juvenile Baboons"), and Nöe and

Sluijter ("Reproductive Tactics") argue that since the males who affiliated with females and their infants were usually in the troop when the infants were conceived, these relationships must reflect paternal investment. This is an odd argument since, prior to these studies, several researchers, including myself, had already shown a significant positive relationship between consorting activity with *particular* females (as opposed to simply presence in the troop) and subsequent male-female and male-infant affiliation (p. 163). These data, along with recent evidence for chacma baboons (Palombit et al., "Friendships"), provide much stronger support for the paternal-investment hypothesis than evidence that the friend was present in the troop when the infant was conceived. However, other results indicate that it is unlikely that *all* affiliation between males and infants reflects paternal investment. First, males occasionally affiliate with infants they could not have sired since they were not present in the troop when the infants were conceived. For example, Nöe and Sluijter ("Reproductive Tactics") report that "some significant [male-infant proximity] scores were found for infants conceived before a male's arrival" (Figure 3 shows five such pairs for one troop and eight for another). Second, many cases exist in which two or three males simultaneously affiliate with the same infants (Altmann, "Baboon Mothers," Stein, "Ontogeny," Nöe and Sluijter, "Reproductive Tactics," Palombit et al., "Friendships"). Since only one of these males could have fathered the infants, the other male (or males) must be investing in someone else's infant. This could result from male inability to accurately estimate paternity, especially when the mother mated with more than one male around the time of conception, but my data suggest that this explanation cannot account for all, or even most, such instances. (Table 8.1 lists nine females with two or more friends [excluding all subadult males] for whom I also had data on mating during the preceding conception cycle. In two of these cases [PH and HH], both male friends were seen consorting with the mother. In the other seven cases, at least one of the friends was never observed consorting with the female, and in one case, neither friend was observed consorting with her.)

As far as I know, only one study since mine has investigated whether females with more than one friend mated with both males during the conception cycle. Palombit et al. ("Friendships") list five females who had two friends and one female who had three friends. For two females, mating information was available for both friends, and in these cases, both friends were seen copulating during the conception cycle. When data from the other females with more than one friend are combined, we find four friends who copulated and five friends for whom data were unavailable. Although the missing data preclude a definitive conclusion, these results suggest that friendship may be more closely tied to prior copulation in this baboon population than in my troop. This result is consistent with other differences between the Okavango chacma baboon population and the Gilgil olive baboon population. Okavango males typically achieve alpha rank soon after migration and quickly monopolize matings with estrous females until another male enters and the process is repeated, usually several months later (Curt D. Busse and William J. Hamilton III, "Infant Carrying by Male Chacma Baboons," *Science*, 212 [1981]: 1281–1283; John B. Bulger, "Dominance Rank and Access to Estrous Females in Male Savanna Baboons," *Behaviour*, 127 [1993]: 67–103; Palombit et al., "Friendships"). In contrast, at Gilgil although prime-aged immigrant males sometimes achieve alpha status right away, high rank does not automatically translate into high consort activity, and immigrants usually experience a fairly long period of integration before they consort frequently (Strum, "Dominance," *Almost Human*, and "Reconciling Aggression"). Thus, after immigration, for at least six months (the gestation period of savanna baboons) Gilgil males can potentially enhance their subsequent consort success by cultivating relationships with females who have infants these males could not have sired. In contrast, for an Okavango male, this same six-month period is likely to offer the best consorting opportunities of his life. It makes sense, then, for Okavango males to concentrate initially on sex and then later, after having sired several infants, to focus more on paternal investment (see also Nöe and Sluijter, "Reproductive Tactics" for yellow baboons at Amboseli).

It seems likely that little consensus will be reached on the benefits to males of friendships with females until DNA analyses routinely yield paternity data for study subjects, and then, I predict, we will find that the selective advantages of friendship vary from one population to the next, as well as among troops, among individuals, and within individuals over time. It may be that for most baboon populations most of the time, males tend to affiliate with females whose infants they could have sired. However, should this prove true, it is important to remember that study after study shows, first, that even the most sexually active males affiliate with some of the females they consorted with and not others (e.g., pp. 161–165), and second, that male-infant affiliation depends upon a male's friendship with the mother. These findings remind us that paternal investment, when it occurs, does not follow invariably from sex per se but rather is mediated by the male's social relationships with females.

8. The labels "fighters," "friends," and "lovers" are mine, but they are based on the male baboon personality types documented by Sapolsky and Ray. See Robert M. Sapolsky and Justina C. Ray, "Styles of Dominance and Their Endocrine Correlates among Wild Olive Baboons *(Papio anubis)*," *American Journal of Primatology*, 18 (1989): 1–13; and Justina C. Ray and Robert M. Sapolsky, "Styles of Male Social Behavior and Their Endocrine Correlates among High-Ranking Wild Baboons," *American Journal of Primatology*, 28 (1992): 231–250. See also Strum, *Almost Human* and "Reconciling Aggression."
9. Strum, *Almost Human* and "Reconciling Aggression." See also my account of Chongo versus Randani (this preface) and Timothy W. Ransom, *Beach Troop of the Gombe* (Lewisburg, Penn.: Bucknell University Press, 1981).
10. Robert M. Sapolsky, "Why Should an Aged Male Baboon Ever Transfer Troops?" *American Journal of Primatology*, 39 (1996): 149–157.

11. Palombit et al., "Friendships."
12. Busse and Hamilton, "Infant Carrying"; Bulger, "Dominance Rank"; Jeanne Altmann, Susan C. Alberts, Susan A. Haines, Jean Dubach, Philip Muruthi, Trevor Coote, Eli Geffen, David J. Cheesman, Raphael S. Mututua, Serah N. Saiyalel, Robert K. Wayne, Robert C. Lacy, and Michael W. Bruford, "Behavior Predicts Genetic Structure in a Wild Primate Group," *Proceedings of the National Academy of Sciences, USA,* 93 (1996): 5797–5801; and Palombit et al., "Friendships."
13. Manzolillo, "Intertroop Transfer"; Strum, "Dominance," *Almost Human,* and "Reconciling Aggression"; and Fred B. Berkovitch, "Reproductive Success in Male Savanna Baboons," *Behavioral Ecology and Sociobiology,* 21 (1987): 163–172.
14. Strum, *Almost Human;* Berkovitch, "Mate Selection"; and Rebecca Dowhan Schneider, "Female Social Preferences and Mating Behavior in Captive Group-Living Baboons *(Papio cynocephalus anubis):* An Experimental Study" (Ph.D. diss., University of Michigan, 1998).
15. Strum, *Almost Human;* Palombit et al., "Friendships"; and Dowhan Schneider, "Female Social Preferences."
16. Rhesus macaques: Bernard Chapais, "Why Do Adult Male and Female Rhesus Monkeys Affiliate during the Birth Season?" in R. Rawlins and M. Kessler, eds., *The Cayo Santiago Macaques: History, Behavior, and Biology* (Albany: State University of New York Press, 1986), pp. 208–217; David A. Hill, "Social Relationships between Adult Male and Female Rhesus Macaques: II. Non-Sexual Affiliative Behavior," *Primates,* 31 (1990): 33–50; and Joseph H. Manson, "Mating Patterns, Mate Choice, and Birth Season Heterosexual Relationships in Free-Ranging Rhesus Macaques," *Primates,* 35 (1994): 417–433. Japanese macaques: this book; Michael A. Huffman, "Mate Selection and Partner Preferences in Female Japanese Macaques," in Linda M. Fedigan and Pamela Asquith, eds., *The Monkeys of Arashiyama. Thirty-Five Years of Study in the East and West* (Albany: State University of New York Press, 1991), pp. 21–53. Chimpanzees: "Special relationships or friendships between males and females have been observed on several occasions at Taï (C. Boesch, unpublished data)," p. 23 in Pascal Gagneux, Christophe Boesch, and David S. Woodruff, "Female Reproductive Strategies, Paternity, and Community Structure in Wild West African Chimpanzees," *Animal Behaviour,* 57 (1999): 19–32. Ringtailed lemurs: Lisa Gould, "Male-Female Affiliative Relationships in Naturally Occurring Ringtailed Lemurs *(Lemur catta)* at the Beza-Mahafaly Reserve, Madagascar," *American Journal of Primatology,* 39 (1996): 63–78. Redfronted lemurs: Michael E. Pereira and Catherine A. McGlynn, "Special Relationships Instead of Female Dominance for Redfronted Lemurs, *Eulemur fulvus rufus,*" *American Journal of Primatology,* 43 (1997): 239–258; and Deborah J. Overdorff, "Are *Eulemur* Species Pair-Bonded? Social Organization and Mating Strategies in *Eulemur fulvus rufus* from 1988–1995 in Southeast Madagascar," *American Journal of Physical Anthropology,* 105 (1998): 153–166.

17. White-faced capuchins: Susan E. Perry, "Male-Female Social Relationships in Wild White-Faced Capuchins, *Cebus capucinus*," *Behaviour,* 134 (1998): 477–510. Wedge-capped capuchins: Timothy G. O'Brien, "Female-Male Social Interactions in Wedge-Capped Capuchin Monkeys: Benefits and Costs of Group Living," *Animal Behaviour,* 41 (1991): 555–567. Woolly spider monkeys (muriquis): Karen B. Strier, "Mate Preferences of Wild Muriqui Monkeys *(Brachyteles arachnoides):* Reproductive and Social Correlates," *Folia Primatologica,* 68 (1997): 120–133 and *Faces in the Forest: The Endangered Muriqui Monkeys of Brazil* (Cambridge, Mass.: Harvard University Press, 1999). Goeldi's monkey: M. H. Jurke, C. R. Pryce, and M. Dobeli, "Sexual Motivation and Behavior in Female Goeldi's Monkey (*Callimico Goeldii*): Effect of Ovarian State, Mate Familiarity and Mate Choice," *American Journal of Primatology,* 36 (1995): 131. Mountain gorillas: David P. Watts, "Social Relationships of Immigrant and Resident Female Mountain Gorillas. I. Male-Female Relationships," *American Journal of Primatology,* 28 (1992): 159–181.
18. Sarah B. Hrdy "Male-Male Competition and Infanticide among the Langurs *(Presbytis entellus)* of Abu, Rajasthan," *Folia Primatologica,* 22 (1974): 19–58, and "Raising Darwin's Consciousness—Female Sexuality and the Prehominid Origins of Patriarchy," *Human Nature,* 8 (1997): 1–49; Barbara B. Smuts and Robert W. Smuts, "Male Aggression and Sexual Coercion of Females in Nonhuman Primates and Other Mammals: Evidence and Theoretical Implications," *Advances in the Study of Behavior,* 22 (1993): 1–63; Richard W. Wrangham and Dale Peterson, *Demonic Males,* (Boston: Houghton Mifflin, 1996); Sarah L. Mesnick, "Sexual Alliances: Evidence and Evolutionary Implications, " in Patricia A. Gowaty, ed., *Feminism and Evolutionary Biology: Boundaries, Intersections and Frontiers* (New York: Chapman and Hall, 1997), pp. 207–260; and Carel P. van Schaik and Peter M. Kappeler, "Infanticide Risk and the Evolution of Male-Female Associations in Primates," *Proceedings of the Royal Society of London,* Series B, 264 (1997): 1687–1694.
19. Barbara B. Smuts and David G. Gubernick, "Male-Infant Relationships in Nonhuman Primates: Paternal Investment or Mating Effort?" in Barry S. Hewlett, ed., *Father-Child Relations: Cultural and Biosocial Contexts* (Hawthorne, New York: Aldine de Gruyter, 1992), pp. 1–30; Carel P. van Schaik and Andreas Paul, "Male Care in Primates: Does It Ever Reflect Paternity?" Evolutionary Anthropology, 7 (1997): 152–156.
20. Examples of primatological research that addresses these kinds of questions include: Jane Goodall, *The Chimpanzees of Gombe: Patterns of Behavior* (Cambridge, Mass.: Harvard University Press, 1986); Altmann, *Baboon Mothers;* Frans B. M. de Waal, *Chimpanzee Politics,* revised edition (Baltimore: Johns Hopkins University Press, 1998), *Peacemaking among Primates* (Cambridge, Mass.: Harvard University Press, 1989), and *Good Natured: The Origins of Right and Wrong in Humans and Other Animals* (Cambridge, Mass.: Harvard University Press, 1996); Strum, *Almost Human;* Dorothy L. Cheney and Robert M. Seyfarth, *How Monkeys See the World: Inside the Mind of Another Species* (Chicago: University of Chicago Press, 1990); Sapol-

sky and Ray, "Styles of Dominance"; and Ray and Sapolsky, "Styles of Male Social Behavior."

21. Kate Baker and Barbara B. Smuts, "Social Relationships among Female Chimpanzees: Diversity between Captive Social Groups," in Richard W. Wrangham, William C. McGrew, Frans B. M. de Waal, and Paul G. Heltne, eds., *Behavioral Diversity in Chimpanzees* (Cambridge, Mass.: Harvard University Press, 1994), pp. 227–242.
22. Rachel A. Smolker, Janet M. Mann, and Barbara B. Smuts, "Use of Signature Whistles during Separations and Reunions by Wild Bottlenose Dolphin Mothers and Infants," *Behavioral Ecology and Sociobiology*, 33 (1993): 393–402; Janet M. Mann and Barbara B. Smuts, "Natal Attraction: Allomaternal Care and Mother-Infant Separations in Wild Bottlenose Dolphins," *Animal Behaviour*, 55 (1998): 1097–1113; and Janet M. Mann and Barbara B. Smuts, "Behavioral Development of Wild Bottlenose Dolphin Infants," *Behaviour* (in press.)
23. Sapolsky and Ray, "Styles of Dominance," and Ray and Sapolsky, "Styles of Male Social Behavior."
24. Fortunately, this entire sequence is captured on videotape.
25. Strum, *Almost Human*.
26. Whenever I mention to people that I visited the Eburru Cliffs troop after a seven-year gap, they ask whether the baboons remembered me. They did. The troop had been observed infrequently but intermittently by a native Kenyan research assistant between 1983 and 1990; otherwise, the baboons were not used to people at close range. When David Gubernick and I approached them, they ran away. I asked David to stay behind and tried approaching on my own. All of the older animals, except for some adult males unfamiliar to me, held their ground. The youngsters (those who would have been born after 1983) ran off initially, but soon calmed down when they saw how comfortable their mothers were around me. In fact, to my astonishment, the females and males I had known in 1983 allowed me to come as close to them as they had at the end of my study seven years earlier (which was very close indeed), and after a glance or two, they seemed as uninterested in me as they were on the day I left. This reveals a remarkable memory for individuals of another species and presumably indicates the existence of at least as good a memory for other baboons.
27. D. Anthony Collins, Curt D. Busse, and Jane Goodall, "Infanticide in Two Populations of Savanna Baboons," in Glenn Hausfater and Sarah B. Hrdy, eds., *Infanticide: Comparative and Evolutionary Perspectives* (Hawthorne, New York: Aldine de Gruyter), pp. 193–215.
28. Randani's babysitting and protection of Hans's infant, including his charge of Chongo, are documented on videotape.
29. Glenn Hausfater and David Takacs, "Structure and Function of Hindquarter Presentations in Yellow Baboons *(Papio cynocephalus)*," *Ethology*, 74 (1987): 297–319; Barbara B. Smuts and John M. Watanabe, "Social Relationships and Ritualized Greetings in Adult Male Olive Baboons *(Papio anubis)*," *International Journal of Primatology*, 11 (1990): 147–172; and J. M. Watanabe and

B. B. Smuts, "Explaining Religion without Explaining It Away: Trust, Truth, and the Evolution of Cooperation in Roy A. Rappaport's 'The Obvious Aspects of Ritual,'" in Aletta Biersack, ed., *Ecologies for Tomorrow: Reading Rappaport Today* (a special issue of *American Anthropologist* in memory of Roy A. Rappaport), in press.

30. Smuts and Watanabe, "Ritualized Greetings," and Watanabe and Smuts, "Explaining Religion."
31. Barbara B. Smuts, "Social Relationships and Life Histories in Primates," in Mary Ellen Morbeck, Alison Galloway, and Adrienne L. Zihlman, eds., *The Evolving Female: A Life-History Perspective* (Princeton: Princeton University Press, 1997), pp. 60–68.
32. James S. Chisholm, "Death, Hope, and Sex: Life History Theory and the Development of Reproductive Strategies," *Current Anthropology,* 34 (1993): 1–24 and "The Evolutionary Ecology of Attachment Organization," *Human Nature* 7 (1996): 1–37; Jay Belsky, "Attachment, Mating, and Parenting: An Evolutionary Interpretation," *Human Nature,* 8 (1997): 361–381; Jennifer N. Davis and Martin Daly, "Evolutionary Theory and the Family," *Quarterly Review of Biology,* 72 (1997): 407–435; Michael E. Kerr, "Bowen Theory and Evolutionary Theory," *Family Systems: A Journal of Natural Systems Thinking in Psychiatry and the Sciences,* 4 (1998): 119–179; Randolph M. Nesse, "Emotional Disorders in Evolutionary Perspective," *British Journal of Medical Psychology,* 71 (1998): 397–415, and "The Evolution of Commitment and the Origins of Religion," *Science and Spirit,* 10 (1999), in press; David M. Buss, *Evolutionary Psychology: The New Science of the Mind* (Boston: Allyn and Bacon, 1999); and April Bleske and David M. Buss, "Can Men and Women Be Just Friends? An Evolutionary Perspective," *Personal Relationships* (under review).
33. Barbara B. Smuts, "Reflections," in Amy Gutmann, ed., *The Lives of Animals* (Princeton: Princeton University Press, 1999), pp. 107–120.
34. Anyone who has been close to a dog or a cat or a similar pet knows what I mean when I talk about personal relationships with individuals from another species. The reason we can form such relationships with dogs and cats is that their ancestors have formed relationships with members of their own kind for millions of years.

Acknowledgments

I am grateful to Robert Smuts, Alice Smuts, and Peter Stine for valuable feedback and to David Gubernick and Anthony Collins for assistance studying the baboons of Gombe.

CONTENTS

5 WHAT MADE FRIENDS SPECIAL

6 BENEFITS OF FRIENDSHIP TO THE FEMALE

7 MALE–MALE COMPETITION FOR MATES

8 BENEFITS OF FRIENDSHIP TO THE MALE

FOREWORD

It is difficult to imagine Africa without the ubiquitous baboon. Surviving all but the most ruthless campaigns of extermination, baboons continue to thrive everywhere, from deep forests to the desert edge. The savannah baboon, the subject of this account, ranges in woodland and veldt from the Tibesti Mountains in the north to Cape Town in the south and across Africa from Djibouti to Dakar. Nowhere else on earth are so many wild mammals of this size living in uneasy commensalism with human populations.

Across the centuries humans have assigned many roles to the baboon. Occasionally, as in Pharaonic Egypt, they held a position of honor in the pantheon. Far more often, they are cast as sly, conniving sneak-thieves, determined to destroy the crops of beleaguered African farmers. Tourists most often see them as temperamental cadgers of tidbits tossed from vehicles. But it is in scientific research that the baboon has found its *métier*.

Over the past 25 years, many hundreds of scientists have accumulated tens of thousands of hours of observations on this absorbing primate, for obvious reasons: they are large, diurnal monkeys that are easy to observe when they forage in short grassland; they are flamboyant in appearance and temperament and highly social by inclination. They are, in fact, nature's gift to the behaviorist, and it is fair to say that baboons (together with their close relatives the macaques) have been as important to field biology as *Drosophila* has been to genetics or the white rat to psychology and medicine.

Those who have been privileged to watch baboons long enough to know them as individuals and who have learned to interpret some of their more subtle interactions will attest that the rapid flow of baboon behavior can at times be overwhelming. In fact, some of the most sophisticated and influential observation methods for sampling vertebrate social behavior grew out of baboon studies, invented by scientists who were trying to cope with the intricacies of baboon behavior. Barbara Smuts' eloquent study of baboons reveals a new depth to their behavior and extends the theories needed to account for it.

Indeed, this volume presages primate studies of the 1980s. While adhering to the most scrupulous methodological strictures, the author yet maintains an open research strategy—respecting her subjects by approaching them with the open mind of an ethnographer and immersing herself in the complexities of baboon social life before formulating her research design, allowing her to detect and document a new level of subtlety in their behavior. At the Gilgil site she could stroll and sit within a few feet of her subjects. By maintaining such proximity she was able to watch and listen to intimate exchanges within the troop; she was able, in other words, to shift the baboons well along the continuum from "subject" to "informant." By doing so she has illuminated new networks of special relationships in baboons. This empirical contribution accompanies theoretical insights that not only help to explain many of the inconsistencies of previous studies but also provide the foundation for a whole new dimension in the study of primate behavior: analysis of the dynamics of long-term, intimate relationships and their evolutionary significance. The importance of this achievement is easier to understand when placed against the earlier history of baboon studies.

Systematic field research on baboons falls into a series of broadly overlapping periods. Early descriptive studies, such as my own, were undertaken in the late 1950s and early 1960s, but these were soon supplanted by the establishment of long-term projects at several sites in which detailed observations by successive teams of scientists made it possible to document the behavior of each individual baboon through most of its lifetime. Correlational analyses of large sets of quantitative data were used to test general hypotheses about social organization and life history variables. The most recent period, beginning in the mid-1970s, was structured by the revolution then taking place in evolutionary vertebrate ecology; newly refined hypotheses were formulated and tested against this new paradigm. Research emphases shifted to questions about the "strategies" in baboon life: inclusive fitness, optimal foraging, resource competition, and life history strategies.

In all of these studies, it was necessary to "objectify" the animals and their behavior in order to purge primate studies of the easy anthropomorphism of the turn of the century. Coding and sampling methods were developed to minimize observer bias, and these still form the backbone of all mammalian studies. By wedding such precise methods to the powerful predictive theories of the new behavioral ecology, scientists believed that the major intellectual questions in baboon studies would soon be answered; what remained was the careful documentation of such principles as kin selection, resource competition, and game theory analyses of aggression. But early returns were puzzling; there were nagging inconsistencies. The most dominant males did not always acquire the most mates. Both males and females exhibited strong mating partner preferences, but the reasons for these preferences were not obvious. Males who were unlikely to have fathered particular infants were nevertheless observed to develop tender, caring relationships with them. And so on. Baboon behavior could not be neatly accommodated by our research methods and theoretical expectations. As this volume so elegantly illustrates, new methods were needed that allowed observers to focus on those behaviors that were most meaningful to the animals themselves, and new theories had to be developed to explain the often surprising findings that emerged.

In retrospect, it is clear that at every stage of research we human observers have underestimated the baboon. These intelligent, curious, emotional, and long-lived creatures are capable of employing stratagems and forming relationships that are not easily detected by traditional research methods. In the process of unravelling their complex social relationships, Smuts has revealed that these masters of strategy and aggressive competition are equally capable of patience, tenderness, and concern. Reading this volume, one feels that the real baboon story is now beginning to unfold.

Irven DeVore
Harvard University

ACKNOWLEDGMENTS

This book reflects the influence of four people who played a central role in my training. Bob Trivers introduced me to evolutionary theory and showed me, by example, the joy of scientific discovery. From Irv DeVore I inherited a fascination with baboons and a conviction that their behavior was complex, multifaceted, and worthy of intensive scrutiny. David Hamburg's wisdom guided me through all phases of my graduate training, and his belief in the evolutionary significance of emotions and enduring social bonds inspired this work. Robert Hinde's insights about how to study relationships helped me to translate my idealistic goals into concrete results.

I watched baboons with several co-workers whose companionship greatly enlivened and enriched my time in the field: Nancy Nicolson, Julie Johnson, Sylvia Howe, and John Watanabe. All four contributed data cited in this book, for which I am very grateful. For helping to make Kenya a home away from home I thank Joab Litzense, Barbie Allen, and the Mehta family of Gilgil, especially Kasim.

Bob Harding and Shirley Strum, codirectors of the Gilgil Baboon Project, made it possible for me to study Eburru Cliffs troop and provided encouragement and support for my research. Dr. Strum introduced me to the baboons and shared her previous knowledge of the troop. During the last phase of this project, Josiah Musau and Francis Milili provided invaluable field assistance under trying circumstances. Robert Sapolsky generously shared information he collected during the capture of the troop's males.

I am grateful to the government and people of Kenya who made this study possible. I thank especially the Office of the President for permission to conduct research and the Gema Cooperative for permission to study baboons on Kekopey Ranch. As it does for all primate researchers in Kenya, the Institute for Primate Research provided me with crucial logistical support. I thank Dr. James G. Else, Director, and the staff of the Institute for their help.

During the long process of writing, Peter Ellison, Paul Harvey, Jim Moore, Shirley Strum, and Richard Wrangham made very useful comments on various drafts of the manuscript. Dorothy Cheney, Robert Seyfarth, Robert Smuts, and John Watanabe meticulously reviewed the entire book and provided many detailed suggestions for its improvement. Without their help this book would have been very different, and I am deeply grateful to them for sharing their time and thoughts so generously. I also thank Jeff Kurland and Frans de Waal for contributing photographs used to open Chapter 10.

The research described in this book was supported by grants to the author from the American Association of University Women, the L.S.B. Leakey Foundation, the W.T. Grant Foundation, and the Wenner-Gren Foundation for Anthropological Research, and by grants to Irven DeVore from the Harry Frank Guggenheim Foundation and the National Science Foundation (#BNS83-03677). Most of the book was written during my tenure as a fellow at the Center for Advanced Study in the Behavioral Sciences at Stanford, California. While at the Center, financial support was provided by the Exxon Educational Foundation and the National Science Foundation (#BNS76-22943). I thank Gardner Lindzey, Muriel Bell, and all members of the Center staff for assistance during my fellowship year.

This book is about the role that friendship plays in the lives of baboons, and it is therefore fitting to conclude by acknowledging the critical role that friendships have played in the completion of this book. During every phase of the work, the colleagues who provided critical appraisals were also the friends I looked to when creative energy ebbed and gumption waned. In addition to those already mentioned, I thank Lila Abu-Lughod, Bob Bailey, Vicki Burbank, Tony Collins, Nancy DeVore, Brazy de Zalduondo, Pippi Ellison, Mitzi Goheen, Sarah Hrdy, Lysa Leland, Kathryn Morris, Laura Norman, Nadine Peacock, Elizabeth Ross, Jon Seger, Karen Strier, Tom Struhsaker, and Pat Whitten. Their unfailing generosity of spirit embodies for me the thesis of this book: friendship indeed lies at the heart of social life.

My family has contributed to this book in so many ways that it is impossible to express fully my appreciation for all they have done. I thank my parents for showing me that the pursuit of challenging work is one of life's main rewards and for supporting all of my efforts to put this lesson into practice. To Mal, Marybeth, and Robert, thank you for being there. Finally, thanks to John Watanabe, for his encouragement, inspiration, and humor throughout every stage of this project, and most of all, for going to Kenya to meet Alex, Daphne, and the others and then telling me, "Now I understand why you love baboons."

He who understands the baboon would do more toward metaphysics than Locke.

Charles Darwin

1 INTRODUCTION

Pandora grooms Virgil on the sleeping cliffs at the end of the day. Pyrrha, Pandora's daughter, peeks out at other baboons resting nearby. Friends often groom on the cliffs in the evening before going to sleep.

PROLOGUE

It was late afternoon, and the angled light brought into clear relief the sculptured planes of the baboons' bony muzzles, highlighting the individuality of each face. In this light, I had no trouble recognizing Virgil and Pandora 100 m away, traveling slightly apart from the rest of the troop, wandering slowly toward me. Their steps were leisurely and their movements relaxed as they picked an occasional handful of young grass or dug for a bulb or root. By this time, the baboons had usually satisfied their voracious appetites, and when they sat they tended to lean back slightly as if to offset the weight of their full stomachs, which bulged above their feet, neatly placed close together. Virgil sat just this way, his chin pointed down and resting against his chest so that his expression, when he glanced up without moving his head, looked a little shy. But there was no hint of shyness in his face when he spotted Pandora, shuffling along behind him, apparently intent on finding a tasty bug or two under the small rocks she was turning over, one by one. He hunched his shoulders, pulled his chin in still further, flattened his ears against his skull, and made the skin around his eyes taut, showing the bright white patches of skin above each eyelid. At the same time, he alternately smacked his lips together rhythmically and grunted deeply with the slight wheeze that distinguished Virgil's voice from those of the other adult males. Pandora, 5 m away, looked up and made a similar face back at Virgil and then, abandoning her rocks, headed toward him with the ungainly trot of a baboon anxious to get somewhere fast, but too lazy to run. As she approached, Virgil lip-smacked and grunted with increasing intensity, as if encouraging her to make haste. When she arrived, she plopped herself down on her back next to him and, dangling one foot in the air, presented her flank in an invitation for grooming. Virgil responded promptly, gently parting the sparse hairs on her belly with his hands, every now and then lightly touching her skin with his lips to remove a bit of dead skin or dirt from her fur.

But Virgil, like most male baboons, preferred being groomed to grooming, and after a few minutes he slowly sank to the ground, expelling his breath in a deep sigh. Pandora groomed him intently, working her way up his neck to the area around his eyes, carefully removing the grit that had accumulated there over the course of the day. After a few moments, they were joined by two of Pandora's offspring, Plutarch, a juvenile male, and Pyrrha, an infant female. Pyrrha was in a rambunctious mood, and she used Virgil's stomach as a trampoline, bouncing up and down with the voiceless chuckles of delight that accompany baboon play. Every now and then Virgil opened his half-shut eyes, peered at Pyrrha, and gently touching her with his index finger he grunted, as if to reassure her that he did not mind the rhythmic impact of her slight body against his full stomach.

After a while, Pandora stopped grooming, and Virgil moved away, slowly clambering up the cliff face where the troop would spend the night. He glanced back every few steps at Pandora and her family, who followed right behind. Finding a good spot halfway up the cliff, Virgil made himself comfortable. Sitting upright, he leaned backward against the rock face, and, grasping his toes in his hands, let his head sink to his chest—a typical baboon sleeping posture. Pandora sat next to him, leaning her body into his, one hand on his knee, her head against his shoulder. Her offspring squeezed in between Pandora and Virgil, and, in the dimming light, I could not tell where the body of one baboon began and the other left off. This is how they would remain for the rest of the night.

Virgil and Pandora were in late middle-age and had lived in Eburru Cliffs (EC) troop for at least 5 years. For the past year, while I had been spending my days with this troop, and probably for long before that, they had been close associates, feeding near one another during the day and sleeping together at night. Although Virgil and Pandora had probably mated in the past, their bond was not dependent on sexual activity. During my time with EC, at first Pandora was pregnant and later nursing her infant, so she was not sexually receptive during this period.

On the cliffs nearby was another pair, Thalia and Alexander. Thalia was an adolescent female who had not yet experienced her first pregnancy. At the moment, she was in the quiescent phase of her monthly sexual cycle, but in a few days the bare skin on her bottom would begin to swell, and for about 2 weeks she would exhibit the exuberant sexuality characteristic of adolescent female baboons. Alexander, also an adolescent, had a long, lanky body and an unusually relaxed disposition. He had transferred into EC troop just a few months earlier. Thalia, like the other females in the troop, was wary of this interloper and tended to avoid him during daily foraging. But as I watched the two of them sitting on the cliffs about 5 m apart, it was clear that, in Thalia, fear and interest were mixed in an uneasy balance.

Alexander was facing west, his sharp muzzle pointing toward the setting sun, watching the rest of the troop make their way up the cliffs. Thalia was grooming herself in a perfunctory manner, her attention elsewhere. Every few seconds she glanced out of the corner of her eye at Alexander without turning her head. Her glances became longer and longer and her grooming more and more desultory until she was staring for long moments at Alexander's profile. Then, as Alexander shifted and turned his head toward Thalia, she snapped her head down and peered intently at her own foot. Alexander looked at her, then away. Thalia stole another glance in his direction, but when he again glanced her way, she resumed her involvement with her foot. For the next 15 minutes, this charade continued: Each time Alexander glanced at Thalia, she feigned indifference, but as soon as he looked away, her gaze was drawn back to his face. Then, without looking at her, Alexander began slowly to edge toward Thalia. Their glances at one another became more frequent, the intervals between them shorter, and their interest in other events less convincing. Finally, Alexander succeeded in catching Thalia's eye as she was turning away. He made a "come-hither" face—the same face Virgil had made at Pandora—grunting as he did so (Figure 1.1). Thalia froze, and for a second she looked into Alexander's eyes. Then, as he began to approach her, she stood, presented her rear to him, and, looking back over her shoulder, darted nervous glances at him. Alexander grasped her hips, lip-smacking wildly, and then presented his side for grooming. Thalia, still nervous, began to groom him. Soon she calmed down, and I found them still together on the cliffs the next morning.

This event represented a triumph for Alexander who, as a newcomer, had been trying for several weeks to establish a relationship with Thalia. As far as I knew, this was not only the first time a female had groomed him for more than a few seconds but also the first time he had spent an entire night close to one. From this moment on, he and Thalia spent more and more time together during the day, until they formed a consistent pair within the troop. Looking back on this event months later, I realized that it marked the beginning of Alexander's integration into the troop.

STUDYING SEX AND FRIENDSHIP IN BABOONS

Soon after I began studying wild baboons in 1976, I realized that relationships like that between Virgil and Pandora or Alexander and Thalia were a central feature of baboon society. The idea of studying these long-term, cross-sex "friendships" provoked my interest for several reasons. Traditional studies of male–female relations in nonhuman primates had focused on male–male competition for mates and on brief bonds between males and sexually receptive females. In many of

Figure 1.1. (*Top*) Adult female with normal facial expression. (*Bottom*) Same female making the "come-hither" face.

these studies, females were treated as relatively passive objects of competition who were of little interest once they conceived and were no longer sexually active.

Because female monkeys are normally either pregnant or lactating, however, periods of sexual activity constitute a very small proportion of their adult lives, and so the vast majority of interactions between males and females from the same troop are not explicitly sexual. Surely these "nonsexual" interactions—whatever they involve—must be crucial to understanding what goes on in the sexual arena. In particular,

I suspected that a detailed knowledge of cross-sex relationships might provide important insights into the role females played in mate selection. Female choice of mates is considered an important force by evolutionary biologists (e.g., Darwin, 1871; Fisher, 1930; Bateman, 1948; Trivers, 1972), but, until recently, primatologists have paid it little attention.

Furthermore, a perspective that included female choice as well as male–male competition might help to explain several puzzling findings. If, as many primatologists had assumed, successful competition against other males was the primary determinant of male access to mates, then high-ranking males consistently should breed more often than lower-ranking ones. This, it turned out, was not always the case; numerous studies reported higher than expected frequencies of mating by lower-ranking, often older, males (DeVore, 1965; Saayman, 1971a; Hausfater, 1975; Packer, 1979b; Manzolillo, 1982; Strum, 1982). Another puzzling finding was the high degree of association between some adult males and particular infants, first documented by Ransom and Ransom (1971). What were these large, ruthless fighters doing in the "female domain," looking slightly out of place as they cuddled and carried tiny infants? Models that focused on aggressive competition among males did not predict the existence of prolonged, affiliative relationships between either males and females or males and infants, yet such relationships clearly existed. I thought that a detailed study of cross-sex friendship might help to explain these findings and also contribute to a more complete understanding of baboon society.

This investigation of baboon friendship involved two major problems: one related to theory and the other to observation. My curiosity about issues like female choice and male–male competition reflects my interest in evolutionary theory, an interest shared by many of the scientists who spend their time watching animals. As intellectual descendants of Darwin, we are all concerned with the problem of ultimate or evolutionary causes of behavior. Has natural selection favored friendships among baboons? Specifically, does having a friend of the opposite sex help an individual to maximize his or her genetic contribution to future generations? These questions assume that male–female friendships are not simply historical accidents or idiosyncratic expressions of baboon psychology, but rather that they are products of evolution analogous to phenomena with more obvious adaptive value, such as maternal care or predator avoidance. This assumption, in turn, allows one to take full advantage of a series of powerful theories for explaining behavior developed by evolutionary biologists over the last 100 years.

This evolutionary perspective played a large role both in formulating my goals at the start of the study and in interpreting results during data analysis. The research itself, however, often took on a very different character due to the powerful influence of my subjects—the baboons. Many primate field workers compare their jobs to watching soap operas, except that the characters are real and they do not speak. This is an apt comparison because what captured my interest and motivated me to return to observe my subjects again and again was the daily drama of baboon life. As a result, all of the larger questions we begin with—for example, how do friendships contribute to individual reproductive success?—quickly become translated into a series of much more immediate, often compelling questions about phenomena of immediate concern to the baboon themselves. How does a friendship form? What makes it survive or, instead, falter in the early stages? Why does this male have half a dozen female friends, while that one has none? Even more basic, what exactly is friendship to a baboon? It was possible that the pairs I considered friends simply represented one end of a continuous distribution, with pairs who spent a lot of time together at one end and pairs who spent much less time together at another. However, it was my impression that interactions between friends differed not only in frequency but also in kind from interactions between other dyads, and that friends did indeed constitute a distinct category of relationships. The challenge was to reformulate this knowledge in a way that would permit both scientific scrutiny and objective communication to others without completely sacrificing the immediacy and vividness of the raw observations. Meeting this methodological challenge was a second major goal of the study.

The organization of this book reflects my two principal theoretical and methodological concerns. Research on any animal must begin with a general knowledge of the species' distribution and social organization, and Chapter 2 provides this information. It also introduces the EC baboon troop and the females and males who were the focus of this study. Chapter 3 describes what it is like to watch baboons and how I structured my observations in order to address the problems and questions discussed above.

Chapter 4 defines baboon friendships and considers how individual attributes like age and status affected the number and types of friends that an individual had. Chapter 5 investigates the role played by each sex in maintaining friendship and the ways in which the interactions of friendly dyads differed from those of other pairs.

Chapters 6, 7, and 8 turn to evolutionary questions by considering how baboon social relationships might contribute to individual reproductive success. Chapter 6 analyzes the reproductive benefits of friendship from the female perspective, including male protection of female friends and affiliative relationships between males and infants. Chapter 7 investigates male–male competition for mates, and Chapter 8 considers the reproductive benefits of friendship from the male perspective. Through an analysis of the intimate connection between sex and friendship, Chapters 7 and 8 attempt to integrate information on male–male competition and female choice into a more unified model of male reproductive strategies.

Chapter 9 turns to some of the questions that to me are the most fascinating and frustrating ones raised by this study. Does the nature of friendship change through the life cycle? Are friendships disrupted by the female's sexual interest in another male? What roles do courtship and possessive behaviors play in establishing and maintaining friendships? Do baboons feel emotions like jealousy, ambivalence, and grief toward their friends?

These are fascinating questions because they concern issues of great relevance to human relationships, but they are also frustrating because they are the most difficult to answer—or even approach—in a scientific manner. Yet because I firmly believe that such questions can and should be addressed without abandoning scientific rigor, I make a preliminary attempt to do so here.

Chapter 10 considers male–female relationships in other primates, including ourselves, in light of issues raised by this study. It concludes by suggesting a different view of the evolution of human pair-bonds from the ones traditionally offered.

Although the first 5 chapters emphasize description and the last 5 interpretation, the two levels of analysis are not really separate. Many of the hypotheses about the evolutionary significance of friendships discussed in the second half of the book are the results of sudden insights that came to me while I was watching baboons. Conversely, the types of behaviors I paid the most attention to in the field were determined, in part, by the evolutionary framework I had adopted. This interplay between theory and observation, which was important at all stages of the research, is reflected in the presentation of information and ideas in this book. My goal in doing so is to convey a feeling for the process, as well as the results, of the study of primate behavior.

2 BABOONS

Eburru Cliffs troop moves away from the sleeping cliffs early in the morning. Some animals, still sleepy, rest quietly while their companions forage in the grass.

INTRODUCTION

Baboons and macaques are grouped together in the subfamily Cercopithecinae, which also includes the less well known mangabeys and guenons. Although there are important exceptions, in general mangabeys and guenons are forest-dwelling monkeys who spend the majority of their time in the trees, whereas baboons (in Africa) and macaques (in Asia) tend to live in more open country, spending much of their time on the ground. Baboons are found throughout sub-Saharan Africa from coast to coast and in the Arabian Peninsula.

Five types of baboons are grouped together in the genus *Papio*: the hamadryas, the Guinea, the yellow, the chacma, and the olive baboons. These five types were originally considered separate species (e.g., Jolly, 1966; Napier and Napier, 1967). Recent evidence, however, indicates that different types interbreed when they come into contact in the wild, suggesting that they are more appropriately viewed as racial variants of a single species, *Papio cynocephalus* (Nagel, 1973; Shotake, 1981; Jolly and Brett, 1973). According to this view, the olive baboons who were the subjects of this study are classified as *Papio cynocephalus anubis*. Throughout, the different types of baboons will be referred to by their common names.

Three of the five types of baboons, the olive, yellow, and chacma, have very similar social organizations. These three types, often grouped together under the name "savannah baboons," live in large groups of anywhere from 20 to 200 animals that contain several adults of both sexes. Savannah baboons are *promiscuous*, a term used by animal behaviorists to describe a breeding system in which both females and males tend to mate with several different members of the opposite sex. Hamadryas baboons, in contrast, are *polygynous*: One male mates exclusively with several females who belong to small, one-male social units. (Little is known about the social organization of Guinea baboons.)

The distribution of savannah baboons is continuous, so that olive baboons, the most northerly species, are replaced by yellow baboons further south; yellow baboons, in turn, are replaced by chacma baboons whose range extends to the tip of southern Africa (Napier and Napier, 1967). Olive baboons, the subjects of this study, are darker, stockier, furrier, and altogether more bearlike than the paler, more gracile yellow baboons, and many chacma baboons are even larger and darker than olive baboons. It is possible that these differences in appearance are also paralleled by differences in social organization and behavior, but our knowledge of the three species is still too imprecise to tell. For this reason and because of their continuous distribution and close genetic relationships (Cronin and Meikle, 1982), observations from all three types of savannah baboons will be treated equally whenever results from different baboon populations are compared.

My study was based on a troop of olive baboons named "Eburru Cliffs," which is part of a population of baboons living near the small town of Gilgil in the Great Rift Valley of Kenya, about 100 km northwest of Nairobi. Further details about the Gilgil baboons and the study troop are given below, but a brief review of the social organization of savannah baboons is given first.

BABOON SOCIAL BEHAVIOR: A BRIEF SUMMARY

Social Organization

Most baboon groups, or "troops" as they are often called, contain several adult males, and because males are almost twice as large as adult females, they tend, at least initially, to capture the observer's attention. Partly for this reason and partly because of androcentric biases, early research on baboons (and many other primates) tended to emphasize the role adult males play in maintaining group cohesion (e.g., Hall and DeVore, 1965). Only after groups were studied for several years did it become obvious that it was the adult females who formed the core of the social system in savannah baboons—a pattern previously described for Japanese macaques (e.g., Kawai, 1958, 1965; Kawamura, 1958, 1965) and rhesus macaques (e.g., Sade, 1965).

In savannah baboons, macaques, vervet monkeys, and most other Old World monkeys, females remain in their natal groups throughout their lives, whereas males generally move to another group as adolescents. The females in a baboon troop are therefore related to one another through common ancestors. Closely related females—mothers and daughters, sisters, and sometimes even grandmothers and granddaughters or aunts and nieces—tend to associate with one another, forming kin-based subgroups within the larger troop. This pattern of association with kin also holds for immature males who have not yet left their natal group (Johnson, 1984). In addition to frequent exchanges of friendly behaviors like grooming and sitting close together, members of the same matriline also support one another during aggressive encounters with other troop members.

Female baboons frequently avoid other females during foraging, and exchanges of threatening and submissive gestures between them are common. Fights also occur, but they are less frequent than threats or submissive behavior. Primatologists use these types of interactions, often referred to as "agonistic behaviors," to determine dominance or "agonistic rank." If, for example, baboon B consistently avoids and shows submission toward baboon A, and A consistently threatens B but does not avoid her, then A is considered dominant to B. Among female baboons, dominance relationships are clear-cut (reversals of outcome are rare), linear (if A ranks above B and B ranks above C, than A also ranks above C, and so on), and relatively stable over many years (Hausfater *et al.*, 1982; but see Smuts, 1980 and in preparation). Adult daughters usually rank just below their mothers, so that whole matrilines can be ranked with reference to one another. These patterns are very similar to those found in macaques (Kawai, 1965; Kawamura, 1965; Sade, 1967; Silk *et al.*, 1981b) and vervets (Seyfarth, 1980; Bramblett *et al.*, 1982).

Many researchers have suggested that in female baboons, macaques, and other Old World monkeys with similar social organization, female dominance rank is an important determinant of female reproductive success, because higher-ranking females often have priority of access to resources that affect reproduction, such as food and water. Sometimes higher-ranking females do score higher on various measures of reproductive success (Dittus, 1979; Dunbar, 1980; Sade *et al.*, 1976; Silk *et al.*, 1981a; Whitten, 1983), but in other cases they do not (Cheney *et al.*, 1981, 1985; Gouzoules *et al.*, 1982). Although long-term effects of female rank are controversial, there is abundant evidence to show that, in the short-term, female rank affects a wide variety of behaviors. For example, high-ranking females are groomed more often by other females (Seyfarth, 1977, 1980) and by immatures (Cheney, 1978; Silk *et al.*, 1981c), are less vulnerable to harassment by other females when they are caring for young infants (Altmann, 1980; Silk, 1980), and have greater access to the best feeding and drinking sites (Cheney *et al.*, 1981; Wrangham, 1981; Whitten, 1983). Because female rank has been shown to have important effects on female behavior, I often include it as a variable in my analyses of male–female interactions.

Although rank and kinship are important determinants of female–female relationships, they are by no means the only significant factors. Unrelated females of disparate ranks sometimes form close bonds (personal observation), and because baboon females are strongly attracted to mothers carrying young infants, all females go through periods of intense social interaction with other females (Seyfarth, 1976; Altmann, 1980).

Unlike females, adult male baboons rarely associate with members of their own sex. Close proximity between adult males is rare, and grooming almost never occurs. With rare exceptions, males come together only to greet, fight, form alliances against other males, or compete over the same resource (Smuts and Watanabe, in preparation). Male dominance relationships are not so clear-cut or stable as those of females (Hausfater, 1975), although in some troops temporary linear hierarchies can be identified (e.g., Packer, 1979b). Male–male relationships are discussed in greater detail in Chapter 7.

Development

Baboon births occur throughout the year, although in some populations there may be seasonal birth peaks (Altmann and Altmann, 1970). Baboons are born with a dark, velvety coat that is replaced by the browner adult pelage by around 7 months. These babies are generally referred to as "black infants." Infants cling to the mother's belly and later ride on her back until they are at least 1 year old, and weaning is completed between 1 and 2 years (Nicolson, 1982). By this time, infants are spending a great deal of time away from the mother, often playing with peers, but they tend to return to her when frightened, during the rest periods that punctuate the day, and at night.

Both females and males reach puberty at about 4–6 years of age, although there is great individual variation in rates of development (Altmann *et al.*, 1977; Scott, 1984). Females continue to grow for only 2 or 3 more years after puberty, but males do not reach full adult size until they are around 9 or 10. Fully grown males have much longer and sharper canines than adult females, the hair around the neck and shoulders is much longer, forming an impressive mane, and they weigh about twice as much as adult females.[1]

By the time growth has stopped, most males have left their natal troop to join another. In the Gilgil population, however, some males remained in their natal troops (Strum, 1982; Manzolillo, 1982; personal observation; see Chapter 7). Partly because males reach full adulthood at a later age than females, the typical baboon troop contains many more adult females than adult males. Mortality rates may also be higher among subadult and adult males than among females of comparable age. It is difficult, however, to obtain accurate information on male mortality, because when a male disappears from a troop, observers often have no way of knowing whether he has died or transferred to another troop.

[1] Mean adult male weight in Eburru Cliffs: 24.4 kg (s.d. 2.4), $N=11$; mean adult female weight: 12.8 kg (s.d. 1.3), $N=30$.

Estrous Cycles and Mating Behavior

Female baboons, like human females, undergo menstrual cycles once they reach puberty, but these cycles are normally referred to as *estrous cycles*, a term applied to the reproductive cycles of many other mammals. Females who are experiencing estrous cycles will be referred to as "cycling females" to differentiate them from *anestrous* (noncycling) females.

Wild female baboons usually have their first estrous cycles when they are 4–6 years old but do not conceive until 1 or 2 years later (Altmann *et al.*, 1977; Strum and Western, 1982; Scott, 1984). Gestation lasts about 6 months. Following the birth of an infant, the mother undergoes a period of amenorrhea when she does not cycle. If the infant dies before being weaned, the mother resumes sexual cycling within a few days or weeks. Otherwise, cycling does not resume for from 5 to 21 months after birth, with a mean of 14 months for EC troop (Nicolson, 1982). Females typically have several cycles before they conceive. The length of the interval spent cycling before conception varies considerably among females of the same troop. It can be as short as 1 month or as long as 15 months; the mean for EC troop was 6 months. In EC, the mean interbirth interval for mothers whose previous infants survived was 26.5 months (Nicolson, 1982).

The estrous cycle lasts for about 35–40 days (Scott, 1984). After 2–3 days of menstruation, the perineal area begins to swell, reaching maximum tumescence within 1–2 weeks. The perineum remains fully swollen for about 7–10 days. A rapid and obvious decrease in the size of the swelling marks the onset of detumescence. The swelling continues to decrease in size for several days and then disappears. About 10 days later, the female menstruates (if she has not conceived), and the cycle begins again. Ovulation occurs about 1–4 days prior to detumescence (Hendrickx and Kraemer, 1969; Wildt *et al.*, 1977; Shaikh *et al.*, 1982). Following Hendrickx and Kraemer (1969), and Hausfater (1975), the day of detumescence is labeled "D-day"; the first day preceding D day is labeled "D-1"; the second preceding day "D-2," and so on. Cycle days that follow the onset of detumescence are labeled "D + 1," "D + 2," etc.

As soon as their perineums begin to swell, adolescent and adult females are sexually receptive, and they will solicit copulations from males of all ages, including infants. Immature males are eager to copulate with any sexually receptive female, but fully adult males are much more discriminating (Hall and DeVore, 1965; Hausfater, 1975; Collins, 1981; Scott, 1984). They show the greatest sexual interest during peak sexual swelling—the time when ovulation is most likely to occur—and less sexual interest when the female's sexual swelling is inflating or deflating. Fully adult males also tend to ignore the first few cycles of adolescent females and the first cycle after postpartum amenorrhea of adult females. These cycles usually do not result in conception (Scott, 1984).

Based on the information presented above, it is estimated that from the time of first conception, female baboons spend about one-fifth (range one-thirtieth to one-third) of their lives cycling. Of this time, only about 1 week per cycle or one-fifth of the cycle length is spent in consort with adult males. Thus, the average female baboon is involved in consortships with adult males for only about one-twenty-fifth of her adult life.

Male–Female Relationships: Evidence from Other Studies

Ransom was the first to describe "intense, long-term relationships or pair bonds between adult males and females" as a regular feature of baboon social life (Ransom, 1981, p. 228; Ransom, 1971; Ransom and Ransom, 1971; Ransom and Rowell, 1972). He discussed these relationships in some detail, focusing on their relevance to the male's relationship with the female's infant. Strum (1975) also described special relationships or "friendships" between adult male and female baboons.

Seyfarth (1978b) provided quantitative evidence for the existence of such special relationships. Using data on proximity, grooming, and agonistic encounters, he showed that two of the eight adult females in a small group had "persistent, high-frequency bonds" with two different adult males. Other females had "low-frequency bonds" with particular males.

The case for the general existence of these bonds was strengthened by publication of J. Altmann's book on baboon mothers and infants (1980). Her quantitative data on proximity and rates of approaches between adult males and females with young infants showed that there was a clear tendency for "particular males to associate with particular mothers" (p. 74). Altmann wrote that these associations reflected "specific relationships between male and female that were established before the infant's birth" (p. 74).

Evidence from these and other studies suggests that at least some aspects of the cross-sex friendships described here for EC troop are widespread, perhaps even universal, in savannah baboon societies. This conclusion is supported by Table 2.1, which lists all the baboon populations in which persistent bonds between adult males and noncycling females have been reported. Reports of long-term relationships between males and females are lacking for only one major study site, Mikumi Park in Tanzania. There is no evidence from Mikumi, however, to indicate that such bonds are absent.

In observational studies, experimental replication of results—a procedure of crucial significance in other scientific domains—is not possible. The nearest thing field primatologists have to replication is the possibility of comparing data from different groups or the same group at different times. Ultimately, this is the only way that we can evaluate and interpret our findings. It is therefore particularly fortunate that data relevant to the issues addressed here are available from other baboon studies, and this information has been included throughout the book.[2] I hope that inclusion of evidence from other studies will help to convince readers that special relationships between male and female baboons are a real and widespread phenomenon. I also hope that this evidence will lead to increased awareness of the variability and complexity of the observations on which field primatology is based, thereby promoting critical appraisal of all generalizations and interpretations, including those contained in this book.

GILGIL BABOONS AND THEIR HABITAT

The Eburru Cliffs troop (named after one of their favorite sleeping sites) is one of at least a dozen baboon troops that inhabit Kekopey cattle ranch located near the town of Gilgil, Kenya. The central part of the ranch, where most observations of baboons have taken place, consists of open grassland studded with occasional patches of bushy shrub (*Tarchonantus camphoratus*), scattered thornbush (*Acacia drepanolobium*), and small groves of giant fever trees (*Acacia xanthophloea*). Subsistence activities of Gilgil baboons have been described in detail by Harding (1973b, 1976). They eat a wide variety of foods including the pods, flowers and gum of the acacia trees, the flowers, leaves, and fruits of bushes and herbs, rhizomes, bulbs, and roots, and, most significant of all, the grass itself. Baboons eat the green blades of young grass during the rainy seasons and dig for corms—the under-

[2] Long-term, special relationships between adult males and females have also been documented in macaques (Takahata, 1982a,b; Chapais, 1983d). Although macaques and baboons share many features of social organization and behavior, they also differ in several important ways that appear to affect male–female relationships (e.g., macaques show less sexual dimorphism in body size than baboons, and they breed seasonally rather than all year round). Because of these differences, comparison of baboon and macaque special relationships will be delayed until Chapter 10.

Table 2.1. Existence of Special Relationships between Adult Males and Females in Different Populations of Savannah Baboons

Study site	Type of baboon	Terms used to describe special relationships	Source
1. Amboseli, Kenya	Yellow	"Godfathers," "associates"	Altmann, 1980
2. Gilgil, Kenya			
a. Pumphouse Gang troop	Olive	"Special relationships," "friendships"	Strum, 1975, 1982; Scott, 1984
b. Eburru Cliffs troop	Olive	"Special relationships," "friendships"	Smuts, 1982; 1983a,b
3. Gombe, Tanzania	Olive	"Pair-bonds"	Ransom, 1971; Ransom and Ransom, 1971; Ransom and Rowell, 1972
4. Masai Mara, Kenya	Olive	"Special relationships," "friendships"	Smuts, unpublished observations
5. Moremi, Botswana	Chacma	"Constant associates"	Busse, 1981; Busse and Hamilton, 1981
6. Mountain Zebra National Park, South Africa	Chacma	"Persistent high-frequency bonds"	Seyfarth, 1978a,b
7. Ruaha, Tanzania	Yellow	"Conspicuous dyadic relationships"	Collins, 1981
8. Suikerbosrand Nature Reserve, S. Africa	Chacma	"Pair-bonds"	Anderson, 1983

ground storage organ of sedge grasses—when the ranch is dry. They also occasionally prey on small mammals such as infant Thompson's gazelles and hares (Harding, 1973a; Strum, 1975, 1981).

Gilgil baboons have large, overlapping home ranges that include a central area of heavy, often exclusive use (Harding, 1976). EC troop wandered over an area of 31 km^2 during the first 2 years of this study. Baboon ranging behavior is affected not only by the distribution of food but also by the location of water and sleeping sites. Rainfall is generally concentrated in two wet seasons, one in November and December and another in April and May. During the wet seasons, Gilgil baboons find water in puddles throughout their range, but when rainfall tapers off, they are dependent on a few water troughs provided for the cattle that have lived on the ranch since the turn of the century.

Baboons are vulnerable to predation by large carnivores, and throughout Africa they achieve some protection from nocturnal predators by sleeping in tall trees or on cliffs. Near Gilgil, cliffs are more abundant than large trees, and every evening a dozen different troops ascend to safety on these rocky outcroppings. Usually each troop sleeps on a different cliff, but occasionally two troops will share a single sleeping site.

In addition to the presence of about 4000 cattle, several other features of the baboons' natural habitat have been altered by human intervention. The number of large feline predators has been drastically reduced through long-standing predator control programs. As a result, lions are rarely seen on Kekopey, but leopards and leopard tracks are occasionally spotted, and we saw cheetahs near the baboons several times. Whenever the baboons spotted cheetahs they alarm-barked, but they did not move away, and the cheetahs ignored the baboons.

Probably the most serious predators of baboons are dogs kept by local poachers, who were occasionally seen hunting on the ranch. EC baboons invariably ran from dogs, heading for the nearest cliffs. During my study, a pack of poacher's dogs caught and killed a pregnant EC female (Aphrodite) as she was running toward a cliff.

Human interference with the baboons was minimal during most of the study period. Whenever the baboons encountered unfamiliar people (such as cattle herders), they simply altered their route so as to avoid them, and the people usually ignored the baboons. There was one exception to this trend: For several weeks in the summer of 1978, the baboons repeatedly raided the barley crop of the only farm located in a heavily used portion of their home range. However, the owners of the crop did not live near their field, and the baboons were chased on only a few occasions when the farmers visited Kekopey. Later, after my study was completed, EC began to raid crops more often, and the frequency of human–baboon interactions increased.

In addition to the cows and the potential predators mentioned previously, the baboons shared their range with antelope such as bushbuck, impala, dik dik, bohor reedbuck, waterbuck, Thompson's gazelles, eland, klip-springer, and steinbok, as well as other large mammals including zebra, warthog, jackals, bat-eared foxes, and African buffalo. Birds of prey were common, but none was seen to prey on baboons.

THE STUDY OF EBURRU CLIFFS TROOP

History

In November 1970, R. S. O. Harding began a 13-month study of the baboon troops on Kekopey ranch, focusing on one troop named the Pumphouse Gang (PHG). PHG has been monitored almost continuously since then by Harding, S. C. Strum, and others. In the fall of 1976, a young adult male born in PHG transferred to EC. Strum, who was engaged in a study of the older natal PHG males, followed this male into EC and observed him there at regular intervals. In April 1977, two more PHG males transferred to EC. The presence of these males, who were used to observers, made it easier for Strum to get close to a new troop, and within a few months she had identified and named all of the EC adult and subadult males, some of the adult females, and a few juveniles.

In August 1977, Nancy Nicolson and I initiated systematic, daily observations on EC. The results described in this book are based on data I collected during three periods: the main study (from September 1977–December 1978) and two shorter follow-up studies (July–September 1979 and May–August 1983). Although the follow-up studies focused on new topics, I continued to monitor male–female relationships. These subsequent observations provide a valuable long-term perspective on cross-sex friendships. Nicolson's study, which focused on mother–infant relations and weaning, also provided information on male–infant relationships that was useful to this study.

Subjects

When Nicolson and I began observing EC, there were 115 baboons in the troop. A breakdown of troop membership by age and sex is shown in Table 2.2. Since I was interested in relationships between sexually mature individuals, all of the 40 adult and adolescent females and 18 adult and subadult males who were present at the start of the study were included as subjects. Table 2.3 shows the dominance ranks, age

Table 2.2. Troop Composition in August 1977 at Start of the Study (N=115)[a]

	Females	Males
Adults	34	14
Adolescents	6	4
Juveniles and infants over 1 year	25	24
Infants under 1 year	4	4
Total	69	46

[a]Adolescent refers to females who had reached menarche but who had not yet given birth to their first offspring and to males who were at least as large as adult females but lacking fully descended canines and full mantle fur (subadult males).

classes, and weights of the female subjects, and Table 2.4 shows ages, residency profiles, and weights of the male subjects. These factors are important because, in the analyses that follow, they are considered as sources of individual differences in behavior.

Weights. In March 1979, shortly after the main study period ended, several members of the Gilgil baboon project and a team of scientists and staff from the Institute of Primate Research, Kenya, trapped most of the members of EC. The baboons were tranquilized, examined, and released later the same day. Weights and other biomedical data were obtained at this time.

Age and Residence. Since female savannah baboons rarely transfer from their natal troops (Packer, 1979a; personal observation), all females were assumed to have been born in EC. Females were assigned to one of five age classes on the basis of physical characteristics that are known to vary consistently with chronological age (Strum and Western, 1982). This procedure is described in detail in Appendix I.

Male ages were more difficult to estimate, because once full size is reached, the relationship between physical characteristics and age is not nearly so regular as it is among females (personal observation; Strum, personal communication). For example, in August 1977, Boz had been a fully adult male for at least 1 year, while Alexander and Sherlock were both subadults approaching full size. When I returned to observe EC in the summer of 1983, both Alex and Sherlock looked like old males: Their canines were heavily worn or broken, their lower jaws sagged, their faces were scarred, and their movements slow and ponderous. Boz, on the other hand, looked and moved very much as he did in 1977.

Due to these difficulties, males were placed into only three age

Table 2.3. Ages, Weights, and Dominance Ranks of Adult and Adolescent Females[a]

Name	Initials[b]	Dominance rank	Age class[c]	Weight (kg)
Dido	DD	1	3	12.4
Zandra	ZD	2	3	12.2
Thalia	TH (a,n)	3	1	10.7
Phaedra	PH (a)	4	1	11.8
Hera	HH	5	2	—
Zizi	ZI	6	3	14.0
Helen	HN (a,n)	7	1	11.1
Phoebe	PO	8	2	13.6
Antigone	AI	9	2	—
Leda	LE	10	2	12.6
Eudora	EU	11	3	13.2
Circe	CI	12	3	15.0
Andromeda	AM (a,n)	13	1	—
Jocasta	JO	14	3	—
Isadora	IS	15	4	13.5
Zena	ZN	16	2	12.8
Lysistrata	LI	17	4	13.8
Thera	TX (n,d)	—	4	—
Clea	CC	18	3	15.0
Rhea	RH	19	2	11.2
Cybelle	CB	20	3	15.2
Delphi	DL (a)	21	1	11.8
Athena	AN (n)	22	5	—
Psyche	PY	23	3	14.6
Medea	MM	24	3	13.2
Aphrodite	AH (n,d)	—	?	—
Cygne	CG	25	2	12.4
Justine	JU	26	3	12.8
Daphne	DP	27	3	13.4
Iolanthe	IO	28	1	11.9
Pallas	PA	29	2	13.8
Olympia	OL	30	2	9.8
Pandora	PA	31	4	10.8
Persephone	PP (n,d)	—	1	—
Melina	ML	32	2	11.8
Xanthe	XA	33	2	12.8
Louise	LU	34	3	13.4
Sophia	SO	35	2	—
Aurora	AU (a)	36	1	12.1
Artemis	AT	37	3	—

[a]See text for discussion of criteria for assigning dominance ranks and age classes and for source of weights. All females who had achieved first pregnancy by the end of the study period are included.

[b]a, Adolescent; d, died during the study; n, not included in analysis of focal animal data (see Chapter 3).

[c]Age classes: 1, young; 2, younger middle-aged; 3, older middle-aged; 4, old. See Appendix I for further details.

Table 2.4. Age, Residence Status, and Weights of Adult and Subadult Males[a]

Name	Initials	Age	Residence status	Date of transfer into EC	Weight (kg)
Agamemnon[b]	AG	Adult	Long-term resident	Unknown	—
Boz	BZ	Adult	Long-term resident	Unknown	26.0
Cyclops[c]	CY	Adult	Long-term resident	Unknown	26.5
Hector	HC	Adult	Long-term resident	Unknown	21.5
Virgil[c]	VR	Adult	Long-term resident	Unknown	25.3
Achilles	AC	Young adult	Short-term resident	Unknown	23.3
Handel	HD	Young adult	Short-term resident	April 1977 (from PHG)	23.0
Ian[b]	IA	Young adult	Short-term resident	April 1977 (from PHG)	—
Sherlock	SK	Young adult	Short-term resident	October 1976 (from PHG)	23.2
Alexander	AA	Young adult	Newcomer	July 1977	29.0
Justinian[b,c]	JS	Adult	Newcomer	June 1977	—
Triton	TN	Young adult	Newcomer	July 1977	25.6
Adonis	AO	Young adult	Presumed natal		21.2
Homer	HM	Young adult	Presumed natal		23.0
Aristotle	AS	Subadult	Presumed natal		—
Hermes	HS	Subadult	Presumed natal		23.0
Plato[b]	PL	Subadult	Presumed natal		—
Pliny	PX	Subadult	Presumed natal		18.8

[a]See text for method of determining ages and residence status.

[b]These males disappeared from EC before the end of the study period. AG, IA, and JS disappeared in August 1978. PL disappeared in October 1978 and was later seen in an adjacent troop.

[c]These males appeared to be considerably older than the other males classified as adults (see text).

classes, which were based on size and growth over a 26-month period (from October 1976, when Strum first observed EC, until the end of the main study period in December 1978). Males who were only slightly larger than adult females in late 1976 were considered subadults. Males who were smaller than fully grown males when the study began but who had reached full size by December 1978 were considered young adults. All males who were fully grown in late 1976 were considered adults. Within the adult group, some males appeared much older than others, and these differences are indicated in Table 2.4, but, for reasons noted above, such impressions may not reflect true ages.

Whether or not a male was born in his troop of residence and, if not, how long he has lived in his current troop are important determinants of male behavior (Packer, 1979a,b; Strum, 1982). Five young adults, (Sherlock, Handel, Ian, Alex, and Triton) and one adult (Justinian) were observed by Strum to transfer into EC, and these males were considered short-term residents or newcomers, depending on how recently they had entered the troop. Newcomers were defined as males who had been in EC for less than 3 months when the study began in August 1977; short-term residents had been in EC from 5 months to about 2 years at the start of the study. For the other males, the residence profiles shown in Table 2.4 represent educated guesses. The four subadults were considered natal residents since most males do not transfer to another troop until they are larger than these four were when they were first identified by Strum. Two of the three young adults present when Strum first observed EC (Adonis and Homer) were also considered natal residents because they each had an extremely close bond with a particular adult female and all of her offspring. Observations of EC in 1983 indicate that such bonds are common between subadult or young adult males and their maternal kin. The other young adult (Achilles) was classified as a short-term resident because his behavior, the behavior of females toward him, and the absence of any close bonds with juveniles suggested that he had probably transferred to EC shortly before Strum's observations began. Finally, the fully adult males already in EC in 1976 were considered long-term residents. Since most baboon males leave their natal troops by the time they are adults, these long-term residents had probably transferred into EC from other troops.[3]

[3] It is possible that one or more of the fully adult, long-term resident males was actually born in EC. When I returned to EC in 1983, three of the seven fully adult males were natal males, and in PHG one natal male never left the

Dominance Ranks. Dominance ranks of adult and adolescent females were based on the outcomes of 3779 dyadic agonistic interactions observed during the study (see Appendix II). Dominance relationships among EC females were linear and completely stable during the main study period. However, between February and July 1979, after the main study had ended, five high-ranking adult females fell dramatically in rank; dominance relationships among all other females remained unchanged. These changes and their repercussions are discussed elsewhere (Smuts, 1980 and in preparation; Johnson, 1984), but it should be noted that relationships with males played no apparent role in these events.

Male dominance ranks were much more difficult to determine because interactions revealing clear-cut rank differences were rare. Therefore, no dominance ranks are given for the males listed in Table 2.4. Dominance and other aspects of male–male relationships are discussed further in Chapter 7.

---

troop (Strum, 1982; Manzolillo, 1982). In EC in 1983, there were few obvious differences between adult natal males and adult transfer males of roughly the same age who had been in the troop for a year or two, in terms of their relationships with adult females. This suggests that for fully adult males who have lived in a troop for several years, lack of information on natal versus transfer status may not be of great importance for the issues investigated in this study.

3 FIELD WORK AND DATA ANALYSIS

The author sharing a shady spot with a few members of Eburru Cliffs troop during a mid-day siesta. To her right is adult male Alex, and behind her, sitting on a branch, subadult male Pan. The others are juveniles; the one immediately to her left is asleep.

HABITUATION

Observers of wild primates must solve two problems before they can begin collecting data: get close enough to the animals to see their behavior clearly and learn to tell individuals apart. When Nancy Nicolson and I began observing EC in August 1977, the baboons were already somewhat used to being followed on foot because of Strum's efforts to observe recent immigrants from PHG. However, many EC members were still quite wary of observers. To reduce their fear, Nicolson and I would stop moving toward the troop as soon as a few baboons began to respond to us, usually by moving away a short distance and glancing repeatedly in our direction. After we stopped moving, they would quickly settle down and ignore our presence. Once the baboons became used to us at a given distance, we would repeat this procedure, and over a period of about 4 weeks the distance at which most individuals responded to us decreased from several hundred to 5 or 10 m. At that point, we were close enough to see and hear our subjects clearly, and it was possible to begin systematic observations.

Since primatologists are interested in how animals behave under natural conditions, we want our subjects to ignore us. The best way to achieve this state is to be as uninteresting as possible; consequently, Nicolson and I never fed the baboons, and when they tried to interact with us we avoided eye contact and slowly moved away. Ignoring the juveniles' invitations to play and acting blasé when confronted with a charging adult male required discipline, but the rewards were great. After about 6 months, most troop members routinely traveled, fed, and rested within a few feet or even inches of us without responding to our presence. As a result, it became possible to observe our subjects from a baboon's perspective—for example, to climb onto the sleeping cliffs with the troop and watch while dozens of baboons settled down for the night all around us. Not only were these exciting experiences but they also made it possible to observe social interactions in fine detail.

RECOGNIZING INDIVIDUALS

When first confronted with the problem of learning to recognize over 100 baboons, the task seemed overwhelming. Except for gross differences in body size, they all looked alike. I began by learning to identify individuals within particular classes of animals, starting with adult males, who were the easiest to spot because of their size. At first, I relied on obvious, highly idiosyncratic markers like a crooked tail or facial scar. These markers were indispensable because they could be used to check whether identifications based on more subtle attributes were correct. Learning practical, reliable sets of cues that would consistently allow me to distinguish one baboon from another required many days of studying different individuals and many long minutes staring at one baboon, waiting for the mental "click" that told me it was one individual rather than another. One day all of the work paid off: I caught a glimpse of a male running past me at full speed 40 m away, and without any conscious effort, I knew who it was. This shift, from recognition based on conscious awareness of particular attributes to one based on unconscious perception of a gestalt, occurred within a couple of weeks for the adults; identification of juveniles and infants took a bit longer. Eventually, like other baboon researchers, I learned to recognize individual baboons in much the same way that I recognize other people.

The other side of this process is the baboons' ability to recognize people individually; if this ability did not exist, habituation would be much more difficult. Baboons learn to recognize observers quickly, and they also seem to rely on gestalt perception, since routine changes in clothing or hairstyle do not interfere. However, radical changes may make a difference: I once left Gilgil for a few days and returned with a dramatically different hairstyle. When I emerged from my car (which the baboons knew well) and began to approach the troop at a distance of about 50 m, the baboons fled. Most of them paused after a few seconds, turned and peered at me intently, and then, one by one, returned to their former activities. Over the next couple of days, I received many wary glances, but they soon became used to my changed appearance.

On several occasions, Nicolson and I tested the baboons' ability to recognize individual people by parking some distance away (up to 100 m) and then having a friend who resembled one of us in size, shape, and clothing emerge and slowly approach the troop while we remained in the car. The baboons invariably responded differently to these strangers from the way they did to us. However, they did not treat all strangers alike. We learned from experience that the more closely strangers resembled familiar observers in appearance and demeanor, the more quickly the baboons came to tolerate their presence. We found it particularly helpful to give a visitor a hat, binoculars, and a clipboard: The baboons seem to have learned that people outfitted in this way are a breed apart from all other humans. They also tolerated strangers accompanied by familiar observers much more readily than new people who approached them on their own. Finally, women evoked less fear than men. We cannot be sure whether this preference for women related to particular characteristics (e.g., women are usually smaller) or to recognition of gender per se. We suspect the latter since, for example, the baboons seemed more nervous around small men than large women.

People often ask me whether the baboons remember observers when they return after long absences. They seem to. When I first approached EC after 7 months away, they exhibited none of the wariness evoked by strangers, and two adults left the main body of the troop, approached within 10 m, and gave me the "come-hither" face described in the introduction. When I returned after nearly 4 years away, many of the individuals I had known before were still present. They immediately seemed very relaxed around me, but I could not be sure whether this was because they remembered me or because they were so used to observers by then that they accepted newcomers with equanimity.

DATA COLLECTION

Daily Routine

EC baboons awake by dawn (around 0645), and they usually leave the cliffs by 0730. The baboons forage throughout the day, and often by 1600 or 1700 they have moved close to the sleeping cliffs where they will spend the night. They begin to ascend the cliffs around 1800, and by nightfall an hour later everyone is settled in for the night.

I usually arrived at the baboons' sleeping cliffs by 0700, and on a typical day I would follow them on foot until I could tell where they were going to spend the night; I often stayed with them until dusk. Once or twice a week I would join them later in the day, again staying with them until nightfall. The advantage of starting the day at dawn was that I could always locate the baboons; when I went out later I sometimes had to spend an hour or two looking for them.

Censusing

On each day spent with the troop, Nicolson and I conducted a census. We recorded the presence or absence of every individual, female reproductive condition, the incidence of new injuries, the condition of previously noted injuries, and signs of illness. For cycling females, I recorded whether the perineum was flat (no swelling), tumescent (swelling increasing in size), fully swollen, or detumescent (swelling decreasing in size). The transition from tumescent to fully swollen was gradual, and subjective assessments at this point in the cycle were unavoidable. However, the decrease in swelling size after the period of full swelling was sudden and could be reliably recorded, and during data analysis it was possible to identify different days of the cycle by counting back from D-day, the first day of detumescence (Hausfater, 1975). Pregnant females could be identified in the field shortly after conception because they ceased sexual cycles and their perineal skin changed over the next few weeks from grey to a deep magenta (the "pregnancy sign"; Altmann, 1970). Finally, lactating females were defined as nursing mothers who had not yet resumed sexual cycling.

Behavioral Observations

The usual goal of a scientist interested in animal behavior is to collect as much information as possible during the time spent with the animals. But two problems arise immediately. First, normally one cannot record everything that every animal does because too many animals are doing too many things at once. This means that decisions must be made about what to record. Second, once the observer decides what to record, further decisions must be made about how to record that information in ways that will minimize biases that interfere with analysis and interpretation of results. Below, my approach to these two problems is briefly described. More detailed information on methods of data collection is provided in Appendixes III, IV, and V.

Deciding What to Record. I began by trying to adopt the attitude of an ethnographer confronted with a previously undescribed society. I made a determined effort to forget everything I knew about how baboons are supposed to behave. Instead, I tried to let the baboons themselves "tell" me what was important. After learning to recognize the animals individually, I spent several weeks observing the troop without recording data systematically. As a result of this period of immersion in baboon society, I developed a number of impressions about male–female relationships. I used these impressions to guide decisions about methods of data collection. To illustrate this process, the following records list four of my strongest impressions and the decisions about data collection that followed from each.

1. Each adult male and female tended to spend time near particular individuals and to groom with these same individuals. These pairs seemed to have special relationships. *Decision*: record spatial proximity of subject to others at regular, predetermined intervals. Also record movements that increase or decrease proximity between subject and others in order to determine which member of the pair is most responsible for maintaining proximity (Hinde and Atkinson, 1970). Record grooming because it is a good indicator of affiliative bonds.

2. Interactions between some male–female pairs were qualitatively different—more relaxed and more prolonged—than interactions be-

tween others. *Decision*: record details of interactions in a way that would allow me to describe objectively the "tone" of an interaction, such as signs of tension (nervous glances, submissive gestures) and signs of ease (engaging in normal activities like feeding while near another). Also record how long the interactions take.

3. Special relationships between females and males appeared to have ramifications in a number of areas, especially in sexual interactions and relations between males and infants. *Decision*: collect information about sexual consortships and male–infant interactions.

4. Cycling females tended to interact with many different males even when they did not show sexual swelling, and males tended to spend more time around these females than around pregnant and lactating females. *Decision*: conduct samples mainly on pregnant and lactating females since long-term relationships will be most apparent in this context. For purposes of comparison, collect some data on cycling females, but consider this comparison to be of lower priority than thorough description of the behavior of pregnant and lactating females.

Several additional decisions about methods of data collection were made based on other impressions. These methods are described in detail in the following chapters.

Deciding How To Record Behaviors. Behaviors can be recorded through either ad lib or focal animal sampling. When using the ad lib method, the observer continuously scans the animals in view and records all observed instances of a given behavior. When using the focal sampling method, the observer concentrates instead on the behavior of one individual at a time and records all instances of several different behaviors. Usually this involves following the animal for a predetermined interval, ranging anywhere from a few mintues to an entire day. The advantages and disadvantages of these two methods are considered carefully in a classic paper by Jeanne Altmann (1974), and I relied heavily on her discussion when making decisions about data collection.

Like most observers, I used both focal and ad lib sampling. The bulk of the behavioral data for the friendship study was derived from focal animal samples, a method that allows unbiased estimates of the rates and durations of activities and behavioral interactions (Altmann, 1974). I sampled focal animals for 30 minutes. All observations were recorded in writing on a checksheet designed for easy retrieval of data. At the start of each sample and at 1-minute intervals for the duration of the sample, I recorded the focal animal's activity (e.g., feeding, resting, grooming; see Appendix V for a complete list of activities) and the identities of all adults and subadults within 5 m of her. At 5-minute intervals, I recorded the identities of all adults within 15 m. These "instantaneous samples" (Altmann, 1974) were used to provide unbiased estimates of time spent in different activities and of time spent near other individuals.

Throughout the focal sample, I recorded every social interaction between the focal animal and all other adult and subadult baboons. I also recorded all interactions between my subject's infant and other adults whenever the infant was within 5 m of the mother. Social interactions were recorded using a shorthand-like code (developed during my 6-month pilot study of another baboon troop in Masai Mara Game Reserve, Kenya). This code allowed rapid recording of fine details of interactions while preserving the sequence of behaviors. These records were used to provide unbiased estimates of the rates and durations of various social interactions. Again, refer to Appendix III for a complete list of the behaviors and types of interactions recorded.

Focal samples were supplemented by ad lib observations. Ad lib sampling was used to record: (1) predation by the baboons; (2) encounters with other baboon troops; (3) any unusual feeding behavior; (4) fights, threats, and chases involving at least one adult; (5) supplants and avoids involving two or more adults (see Appendix III for definitions); (6) prolonged proximity, grooming, and other likely indicators of kinship between adult females and other troop members; (7) grooming between adults; and (8) sexual consort partners. This list includes both uncommon behaviors that were rarely observed during focal samples (e.g., predations) and behaviors that provided useful background data for the friendship study (e.g., sexual consort partners).

Choice of Focal Animals

Although I was interested in the behavior of both adult males and females, most of my focal samples were conducted on females only for the following reason. Even though adult females are in sexual consortships with adult males only about one-twenty-fifth of the time, in a troop with 40 mature females, at any given time a number of females will be cycling, and several will have sexual swellings. Males tend to focus their attention on these sexually receptive females. Thus, focal samples on males, rather than on females, would result in more information about sexual competition for mates but less information on male interactions with anestrous females. Since several previous studies had focused on the former, and since I was primarily interested in the latter, I sampled mainly females and only females who were not in sexual consortships. Information on consort partners was derived from daily ad lib records of sexual consortships.

Near the end of the study, I was joined by a research assistant, Sylvia Howe, and we collected focal samples on the 14 remaining adult and subadult males from October through December 1978. Because these samples covered only a brief period, I have not used them in most of the analyses described in this book, but data from male focal samples were included in analyses of male–female aggression (Chapter 6) and male–male relationships (Chapter 7).

All 34 adult females in the troop at the start of the study were included as focal animals (see Table 2.3). An adult female was defined as any female who had given birth to at least one offspring. Adolescent females, defined as females who had begun sexual cycling but who had not yet given birth for the first time, were not sampled. When an adolescent female gave birth and her infant survived, she was added to the list of subjects. Three females (PH, DL, and AU) made this transition during the study period. Three adult females died early in the study period (AH, PP, and TX) and were not included in the analyses reported here. A fourth female, AN, periodically disappeared for days or weeks at a time, making it impossible to sample her behavior on a regular basis. She was a very old female who had ceased cycling and interacted little with other members of her troop. Due to her unusual behavior and her small number of focal samples, she was also eliminated from the analyses. This left 33 females as focal animals.

Sampling of these females resulted in almost 1000 hours of focal animal data. However, because the frequency of sampling depended on female reproductive condition, different females contributed variable numbers of focal samples to this total (see Appendix IV). I have resolved this bias by treating data from each female as a separate data point for most analyses rather than combining results across females.

The day was divided into six 2-hour time blocks beginning at 0700 and ending at 1900. Focal samples for each female were distributed nearly evenly over these time blocks, with slightly fewer samples in the first and last intervals. This procedure minimized biases related to time of day. Subjects were chosen on a semirandom basis by sampling the first female seen from among those females who had been sampled less often than other females during a given time block. Appendix IV lists the total sampling time and number of samples obtained for each female for each reproductive condition.

TYPES OF EVIDENCE

Partly as a result of decisions I made about data collection and partly as a result of the frequency and observability of different types of behaviors, by the end of the study I had data of widely varying quality. The data fell into three main categories:

1. *Quantitative Data Susceptible to Statistical Analysis.* Examples include focal animal data on spatial proximity and on male–female interactions that occurred relatively often (e.g., supplants) and ad lib data on grooming partners and consort activity. It was possible to analyze these data using standard statistical techniques. Statistics are used both to support descriptive statements (e.g., most females groom primarily with one or two males) and to test hypotheses (e.g., females tend to associate with males with whom they consorted most frequently in the past).

2. *Quantitative Data of Insufficient Sample Sizes for Statistical Analysis.* Examples include affiliative interactions between males and infants, some types of male–female interactions (e.g., male aggression), and male carrying of infants in agonistic encounters with other males. These behaviors were recorded systematically, but they were uncommon. Because of small sample sizes, statistical analysis was not possible, but the data often indicated interesting trends. In these cases, I report the results in a quantitative format and use the trends to generate hypotheses that may be worth further investigation.

3. *Qualitative Data.* This category includes impressions and interpretations based either on a series of descriptive observations or on single events that seemed particularly informative. I use such qualitative information to suggest or support an explanation for other results, to illustrate generalizations, and to generate hypotheses. Clearly, information of this kind does not constitute scientific evidence: The information is descriptive rather than quantitative and therefore liable to subjective biases, and my choice of which anecdotes and impressions to include adds still another subjective element to the procedure. These impressions, however, are a valuable source of interpretations and hypotheses that can be subjected to empirical verification, and they play an important role in scientific research. In many places throughout the text, I have inserted anecdotes, highlighting them by the use of a different type face. My aim is twofold: first, to distinguish these observations explicitly from scientific evidence and, second, to encourage readers to share the dynamic interplay between quantitative anlysis and more intuitive mental processes that occurs in the mind of the scientist.

PRESENTATION OF RESULTS

I expect that many of the people who read this book will be fellow scientists, including other primatologists, some of whom will be interested in critically evaluating the interpretations and conclusions I present. Such evaluations are possible only when detailed information about the data and the analytical methods employed is made available. On the other hand, I also hope this book will be of interest to laypersons and people in other fields. Many of these readers will no doubt be more interested in the substance of my findings than in the details of the methodology. To resolve the conflicting needs of these different types of readers I have adopted the following conventions. First, in the text I have included a relatively small number of tables to present and quantify some of the major results, but in many cases I have relegated tables and explanatory notes to the Appendixes. This allows readers who want more details to find them easily without cluttering the text. Second, I have indicated the use of statistical tests by numbers in square brackets in the text. The corresponding statistical tests, sample sizes, and significance levels are listed at the end of each chapter. When I state in the text that two samples differ significantly, this indicates a probability of .05 or less. Unless stated otherwise, two-tailed probabilities were used to test for significance.

4 DEFINING FRIENDSHIP

Cyclops' sprawled posture and half-closed eyes indicate his pleasure in being groomed by his Friend, Zizi. Frequent, relaxed grooming is one of the distinguishing characteristics of friendship.

INTRODUCTION

I returned from the field with a suitcase full of data sheets and a number of questions about baboon friendship that I was eager to explore. My first task was to devise an objective definition of friendship. The significance of this step may be clarified by a brief foray into laboratory science. In the lab, an experiment begins with a careful determination of one or more "independent variables." These variables provide the experimenter with two or more groups that differ in important respects, and during the experiment these groups are compared in terms of one or more measures, referred to as "dependent variables." If the groups differ significantly in terms of these dependent variables, the experiment has indicated that the independent variables have notable effects and that they are therefore important. For example, in a medical experiment the independent variables might be different drug treatments and the dependent variables the effects of these treatments on the course of a disease. In a behavioral experiment, the independent variable might be kinship (the subjects would be divided into two groups, one containing close relatives and the other unrelated individuals), and a dependent variable might be willingness to share food provided by the experimenter. In this case, the results would help to determine whether kinship is an important factor to consider when trying to understand food-sharing behavior.

My goal in the friendship study was to determine whether the type of long-term relationship a male and female had (the independent variable) was an important factor to consider when trying to understand behaviors like male protection of infants and female choice of mates (dependent variables). To achieve this goal, it was necessary to classify male–female relationships into different types that could then be compared systematically with reference to these and other dependent variables. Finding behavioral criteria that would differentiate some male–female relationships from others was therefore the crucial first step of data analysis.

This step entailed several smaller steps: (1) deciding which measures to employ to generate behavioral criteria for defining friendship; (2) examining the distribution of these measures across all male–female dyads; (3) comparing these distributions for each measure; and (4) using these comparisons to define friendship. The results of these four steps are the subject of this chapter.

BEHAVIORS USED TO DEFINE FRIENDSHIP: PROXIMITY AND GROOMING

From the earliest days of primatology, researchers have used spatial proximity and grooming as measures of affinity between individuals (e.g., Carpenter, 1963, p. 36, Washburn and DeVore, 1963, p. 107). The general usefulness of these two measures is illustrated by numerous studies showing that the frequency of grooming or proximity (or both) is related consistently to other important variables such as biological kinship (e.g., Sade, 1965), the frequency of alliance formation (e.g., Cheney, 1977a; Dunbar, 1980), the frequency of fighting, mounting, and presenting (e.g., Kummer *et al.*, 1978), or the probability of future copulations (e.g., Michael *et al.*, 1978). It is also likely that primates themselves use the frequency of proximity and grooming to judge the strength of a relationship between other members of their group (e.g., Bachmann and Kummer, 1980), although this has yet to be conclusively demonstrated. For these reasons, I decided to use grooming and proximity to measure affinity between adult male and female baboons.

GROOMING

Most baboon grooming occurs early in the morning while the animals are still on the sleeping cliffs, during daytime "siestas" when the troop relaxes near a waterhole or in a shady spot, and in the late afternoon as the animals begin leisurely to ascend the sleeping cliffs. On occasion a pair or small cluster of baboons settles down to groom while the rest of the troop forages around them. Most grooming episodes last 5–10 minutes although they can be as brief as a few seconds or as long as an hour. All combinations of age/sex classes groom together except adult males, but the most common grooming dyads involve relatives, juvenile or adult females and a lactating mother, and adult males and females. Grooming appears to be pleasurable for both partners. The groomer lip-smacks and grunts softly while parting the partner's fur and gently removing dirt and ectoparasites, and the groomee tends to adopt a relaxed posture, sometimes sprawling on the ground with limbs splayed and eyes closed.

Throughout the day, I recorded grooming between females and males on an ad lib basis. Each grooming record indicated the time and the identities of the groomer and groomee. If I observed a change in the direction of grooming after a grooming pair was first observed, this was also noted. In order to ensure independence between records, the same grooming pair was recorded a second time only if the end of one grooming episode and the start of another were separated by at least 15 minutes and the pair was observed in a new location. In practice, grooming between the same partners on the same day usually occurred several hours apart.[1]

By the end of the main study, I had collected over 1000 ad lib observations of grooming between males and females. I separated these records into three groups: those involving pregnant or lactating (i.e., anestrous) females, those involving cycling females who were not in consortship, and those involving consort partners. I eliminated the last group from analysis because when in consort the female is involved in a short-term, exclusive, sexual relationship that radically alters her normal pattern of interaction with males (Hausfater, 1975; Rasmussen, 1980). The records involving anestrous females and those involving cycling females not in consort were analyzed separately to see whether female reproductive condition had any effect on male–female grooming interactions.

Do Anestrous Females Have Favorite Male Grooming Partners?

For each anestrous female, I determined the percentage of all of her grooming episodes with males that involved each of the 18 male subjects[2] (groomer/groomee roles were ignored in this analysis, but

[1] I used ad lib data for the grooming analysis because they provided a much larger sample of grooming bouts than did focal samples. I minimized possible biases against less conspicuous grooming pairs (e.g., those who tended to travel on the periphery of the troop) by making a special effort to move through the entire troop and to record all adult grooming pairs whenever the animals were resting. Now and then, while most of the troop was foraging and I was engaged in focal sampling, a few pairs would be observed grooming. Biases introduced by spatial factors were perhaps greater under these circumstances. However, because the troop spent most of its time in open country, even individuals on the edge of the troop were usually easily observed and identified, and it therefore seems likely that the ad lib grooming records represent a fairly good approximation of the actual relative frequencies of grooming between members of different male–female pairs.

[2] The number of males available as grooming partners varied from 18 at the start of the study period to 14 at the end. Three adult males (AG, IA, and JS) disappeared in August 1978, nearly two-thirds of the way through the study, and a subadult male (PL) disappeared in October 1978, near the end of the study. No males joined the troop during the study. Among pregnant and lactating females, only four groomed with a male who was not present throughout the study on more than 5% of her grooming episodes (AI with AG, LE with AG, AT with AG, and CG with IA). AI and AT both resumed cycling just after AG disappeared so that he was present in the troop during the entire time they were pregnant or lactating. For the other two females, LE and CG, ad lib grooming data were analyzed separately for the periods before (period 1) and after (period 2) the males' disappearance.

most episodes involved females grooming males; see Chapter 5). Of the 36 adult and adolescent females, 34 showed a significant preference for one male grooming partner, and one-half of the females also showed a significant preference for a second male partner [*1*] (see Table 4.1). (The two females who did not show a preference, SO and XA, were hardly ever seen grooming with males.) The male with whom the female groomed the most often was termed her primary grooming partner or "P" male. Other males with whom she groomed in at least 20% of all grooming episodes with males were termed secondary grooming partners. A glance at Table 4.1 shows that the identities of these favorite partners varied across females.

To what extent did females restrict their grooming to one or two males? To answer this question, I combined data for all females and examined the cumulative number of grooming episodes with P males, males with the second highest score, and so on (Figure 4.1). All females combined performed 66% of all grooming with males with the P male (range 32–100%), and 86% of all grooming with the first- and second-ranked males combined (range 60–100%). In other words, the vast majority of an anestrous female's grooming with males was typically restricted to 1 or 2 of the 18 possible partners.

Similar results emerged when the favorite female grooming partners of males were considered; only anestrous females were included in this analysis (Table 4.2). All but one male (who was seen grooming with a female only once) exhibited a significant preference for one P female, and one-half also showed a significant preference for a secondary partner [*2*]. All males combined performed 39% of all grooming with the P female (range 23–100%), and 62% of all grooming with the first- and second-ranked female partners (range 44–100%). The numbers alone suggest that males were somewhat less discriminating in their "choice" of grooming partners than were females, but this may simply be a result of the greater number of females available as grooming partners.

Cycling Females Compared with Anestrous Females

While pregnant and lactating females tended to associate mainly with relatives, male friends, and other mothers or mothers-to-be, cycling females seemed to be less selective, interacting with a wide variety of partners including many different males. If this impression were accurate, one would not expect cycling females to show as strong a tendency to restrict grooming to one or two particular males as did anestrous females. To test this prediction, I calculated for each cycling female the percentage of all grooming with males outside of sexual consortships that involved each of the 18 male subjects, as in the

Table 4.1. Males with Whom Anestrous Females Groomed for at Least 20% of All Grooming Episodes with Males[a]

Female	Male	Grooming (%)	P_m	Male	Grooming (%)	P_m	Male	Grooming (%)	P_m	Total no. of grooming episodes with males
AI	AG	50.0	***	SK	37.5	*				8
AM	BZ	52.2	***	AO	34.8	***				23
AT	HS	32.0	***	TN	32.0	***	AG	20.0	***	25
AU	AS	73.1	***							26
CB	HD	42.9	***	HC	28.6	*				14
CC	CY	37.5	*	HM	25.0	ns				8
CG(1)[b]	IA	85.7	***							14
CG(2)	SK	50.0	***	AS	37.5	*				8
CI	HC	77.8	***	VR	22.5	ns				9
DD	HD	75.0	***							28
DL	HM	39.4	***	PX	39.4	***				33
DP	PX	46.2	***	HD	28.2	***				39
EU	VR	48.8	***	HC	34.2	***				41
HH	SK	55.9	***	BZ	23.5	**				34
HN	SK	38.1	***	AO	23.8	*				21
IO	AS	70.0	***							30
IS	VR	69.2	***	HC	30.7	***				26
JO	HC	95.0	***							20
JU	SK	71.4	***							21
LE(1)[b]	AC	47.1	***	AG	29.4	***				17
LE(2)	AC	55.0	***	AS	30.0	**				20
LI	HM	62.8	***	HD	27.9	***				43
LU	VR	100.0	***							19
ML	PX	45.0	***							20
MM	AC	85.7	***							14
OL	AC	44.4	**	AS	44.4	**				9
PA	VR	100.0	***							14
PH	AO	95.0	***							20
PO	CY	83.3	***							60
PS	AC	50.0	***							6
PY	HD	91.3	***							23
RH	BZ	91.7	***							36
TH	AA	89.5	***							19
ZD	AO	54.2	***							29
ZI	CY	54.2	***	SK	45.8	***				24
ZN	TN	40.0	***	AO	33.3	**				15
Total										816

[a] P_m = the probability that a female would groom with the male as often as she did based on the null hypothesis of no partner preferences. See [*1*] for method of computing P_m. *** < .001, ** < .01, * < .05.

[b] Data for CG and LE are divided into two periods: (1) before and (2) after a P partner disappeared from the troop.

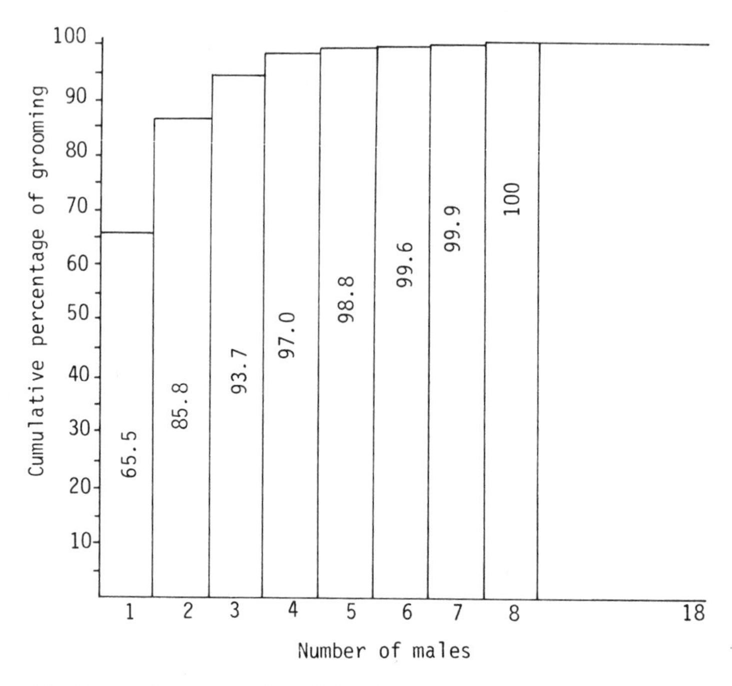

Figure 4.1. Cumulative proportion of female ad lib grooming with males: anestrous females. The number of each female's ad lib grooming episodes with the male with whom she groomed most often (1), second most often (2), and so on, was determined. Then, for each "category" of male (1,2,3, etc.), these values were summed across all 34 females. The proportion of the total number of grooming episodes involving anestrous females and males ($N = 816$) was calculated for each category of male and then added to the combined proportion of all smaller-numbered categories to produce a cumulative frequency distribution.

preceding analysis. Two results supported my expectations. First, for 18 of 22 cycling females, the P male accounted for a smaller proportion of grooming than when the same female was anestrus [*3*]. Second, the distribution of the cumulative amount of grooming accounted for by different numbers of males (see Figure 4.2) differed significantly from the distribution for anestrous females [*4*]. This meant that the same number of males accounted for a smaller percentage of total grooming episodes for cycling females compared with anestrous females. For example, among cycling females the first and second favorite grooming partners together accounted for only 65% of all grooming, compared with 86% for anestrous females. When the data were viewed from the male perspective, the results were similar: The distribution of the cumulative amount of grooming accounted for by different numbers of

Table 4.2. Females with Whom Males Groomed for at Least 20% of All Grooming Episodes with Anestrous Females[a]

Male	Female	Grooming (%)	P_f	Female	Grooming (%)	P_f	Female	Grooming (%)	P_f	Total no. of grooming episodes with females
AA	TH	94.4	***							18
AC	LE	41.3	***							46
AG	AT	22.7	***	LE	22.7	***				22
AO	PH	33.9	***	ZD	32.1	***				61
BZ	RH	50.0	***							66
CY	PO	65.8	***							76
HC	JO	28.9	***	EU	21.2	***				66
HD	DD	27.3	***	PY	27.3	***				77
HM	LI	61.4	***	DL	30.0	***				43
IA	CG	100.0	***							12
JS	—									1
SK	HH	26.4	***	JU	20.8	***				72
TN	AT	42.1	***							19
VR	EU	22.7	***	LU	21.6	***	IS	20.5	***	87
AS[b]	IO	28.0	***	AU	25.3	***				68
PX[b]	DP	34.6	***	DL	25.0					52
PL[b]	IO	23.5	***							15
HS[b]	AT	47.1	***							15
Total										816

[a]P_m = the probability that a male would groom with the female as often as he did based on the null hypothesis of no partner preferences. *** < .001. See [2].

[b]These four were subadult males.

cycling females differed significantly from the distribution for anestrous females [5]. When males groomed with cycling females, the first- and second-ranked females accounted for only 52% of all grooming compared with 62% for anestrous females.

Thus, when a female was cycling, she tended to distribute her grooming with males over more individuals and to groom with the P male for a smaller percentage of her grooming episodes. Similarly, males showed a reduced tendency to groom with particular females when their grooming scores for cycling females were compared with their scores for anestrous females. However, many cycling females (15 of 22, or 68%) continued to show a significant preference for one

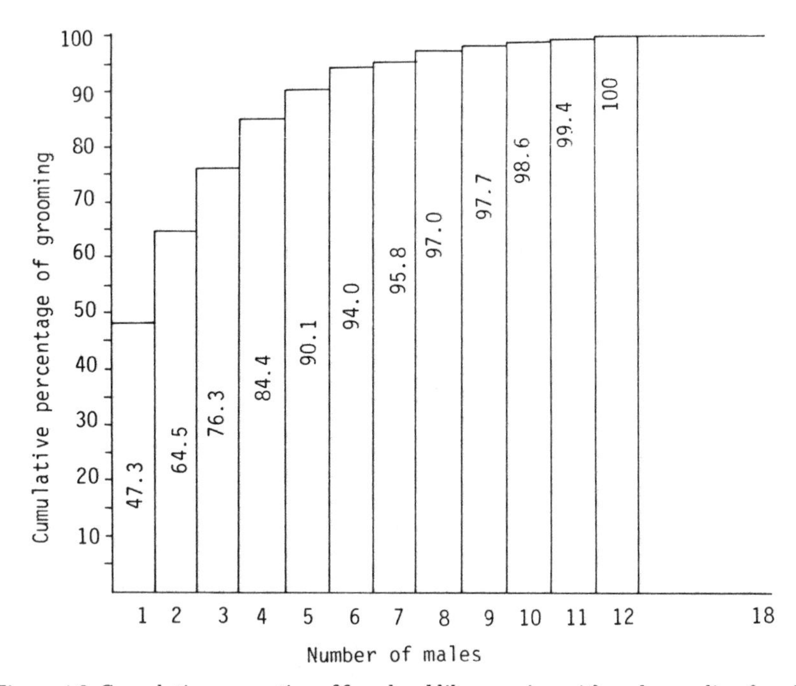

Figure 4.2. Cumulative proportion of female ad lib grooming with males: cycling females. The cumulative frequency distribution was determined as in Figure 4.1, using ad lib grooming data for cycling females (number of grooming episodes with males = 262). Twenty-two different females contributed data.

P male, and two also showed a significant preference for a secondary grooming partner [*6*] (it is likely that other cycling females also had preferences for secondary partners, but due to the smaller number of observations for cycling females this was difficult to detect). Similarly, 8 of the 12 males who were observed grooming with cycling females often enough to be included in the analysis showed a significant preference for one partner, and 4 of these males also showed a significant preference for a secondary grooming partner [*7*].

Degree of Female Preference for Primary Grooming Partner: Anestrous Females

The amount of grooming with the P male by anestrous females ranged from 32 to 100%, and I wondered whether this variation was related to female characteristics such as dominance rank or age. However, neither factor appeared to play a role [*8, 9*]. Old females appeared to groom P males slightly more, but the small sample size made it impossible to tell whether this trend represented a genuine difference.

We would expect females who had only one main grooming partner to show a higher proportion of grooming with the P male than the females who had two or three favorite grooming partners. This hypothesis was supported: The 17 females who had more than one favorite grooming partner (see Table 4.1) groomed the P male significantly less (mean 50%) than the 16 females who had only a P partner (mean 77%) [*10*]. Since these two groups of females did not differ in dominance rank or age, however, we are still left with the question of why some females restricted themselves to one male.

Another factor that might be relevant is the number of females who shared a P male, since competition for the male might reduce a female's access to him (e.g., Seyfarth, 1978b). However, this hypothesis was not supported by the data: There was no tendency for females who shared their P partner with other females to groom with him less. In fact, the two females (LU and PA) who groomed with their P partner 100% of the time had the same partner (VR).

A more subtle version of the same hypothesis suggests that if two or more females shared a P male, only the lower-ranking female(s) would show a reduction in the proportion of time spent grooming. But again, there was no tendency for the lower-ranking partners to groom with the P male less. For example, both LU and PA were lower-ranking than VR's two other P partners who groomed with him less than did LU and PA.

The tendency for some females to restrict their affiliative behavior to one male and for others to associate with two or three males is also apparent in the proximity data (next section). The fact that neither age nor dominance rank distinguishes these two groups is consistent with my impression that the determinants of affiliative relationships between adult males and females are multiple and complex. It is likely that many aspects of male–female relationships can only be understood in terms of the long-term histories of the individuals involved. Qualitative evidence in support of this view is presented in Chapter 9.

SPATIAL PROXIMITY BETWEEN ADULT FEMALES AND MALES

Although the results of the analysis of ad lib grooming records were clear, it was important to test whether the tendency to affiliate with particular partners was also apparent in observations collected during focal samples—the ideal method for providing unbiased estimates of behavioral frequencies. As noted in Chapter 2, during each focal female sample I recorded the identities of all adults within 15 m of the subject at 5-minute intervals (instantaneous samples, Altmann, 1974). Males were recorded in one of the following distance categories: one (0–1 m), two (1–2 m), three (2–5 m), or four (5–15 m).

For each distance category, I obtained an estimate of the percentage of time a female spent in proximity to an adult male by dividing the number of instantaneous samples in which the male was present at that distance by the total number of instantaneous samples for the female. For all females, records were tallied separately for the anestrous and cycling conditions. Only data for the anestrous condition are discussed in this section.

Each anestrous female, then, had four sets of proximity estimates, or scores, for each male. Clearly, scores for different distance categories are not independent and should not be analyzed separately. However, the use of a more gross measure, such as percentage of time spent within 15 m, would result in the loss of information about proximity over the smaller distances.

A method that combined information from all four scores therefore was needed. When the four distance categories are pictured as a series of concentric circles with the female at the center, it becomes clear that the area within which a neighbor could be found was greater the larger the distance category, and so the percentage of time spent near a male would be consistently greater the larger the distance category. If the scores for different distance categories were simply added together, those from the larger categories would carry much more weight and would tend to swamp the contributions of scores from the smaller categories. Therefore, each set of scores was multiplied by a factor, X_D, where D = distance category and $X_1 > X_2 > X_3 > X_4$. These weighted scores were then summed to form a "composite" proximity score (C score). (The weighting factors and an illustration of the method of calculating a male/female pair's C scores are shown in Appendixes VI and VII.)

For each female, I plotted the distribution of the C scores of all 18 adult and subadult males; examples for four typical females are shown in Figure 4.3. If the female associated with males at random, we would expect the distribution of C scores to be more or less continuous. If, on the other hand, the female tended to associate with particular males, the distribution should be discontinuous: One or a few males should have much higher C scores than all of the others. In general, this second pattern held. For all but three females, there were one, two, or three males whose *C* score(s) were separated from the *C* scores of all other males by at least one interval on the scale of measurement (the y axis in Figure 4.3). On average, the *C* score of this male (or males) was separated by 3.6 intervals from the scores of other males, and two females had one male with a *C* score 13 intervals higher than any other male. These results support the hypothesis that most anestrous females associated primarily with one, two, or occasionally three males.

COMPARISON OF GROOMING AND PROXIMITY

For most females, the primary associates revealed by the proximity analysis were the same males with whom they groomed most often. This overlap is illustrated in Figure 4.3, which identifies the females' primary and secondary grooming partners with solid and cross-hatched bars, respectively. For 27 of the 29 anestrous females (93%), the P partner ranked either 1 or 2 on *C* score, and for 16 of the 20 females (80%) who had secondary grooming partners, that partner ranked either 1 or 2 on *C* score. If we turn the comparison around and instead determine how the males who ranked 1 or 2 on *C* scores ranked on ad lib grooming, the results are very similar.

It might be argued that the similarity between the identities of male grooming partners and males in proximity is to be expected, because females will simply tend to groom with those males who happen to be near them. This explanation for the concurrence between grooming and proximity scores is unlikely for two reasons. First, females did not always have high grooming scores with males who were frequently found near them; this indicates that proximity by itself is inadequate to stimulate a grooming episode. Quantitative support for this argument is provided by Collins (1981). Second, daytime rest periods and late afternoon socializing—the times when most of my grooming observations occurred—were generally preceded by a great deal of movement and resultant resorting of neighbors, so that the animals closest to an individual at these times were often not the same ones she or he was near a few minutes before while foraging. Juveniles and older infants provide one of the best illustrations of this principle. Immature offspring are often away from the mother for several hours at a time, playing with peers and foraging independently. Yet, as soon as a female settles down to rest, her immature offspring tend to materialize by her side as if by magic. Often, these sudden appearances are followed by grooming between mother and offspring. The same is often true of adult males and females: When the troop begins to settle down to rest, a male or female who has not been near a potential grooming partner for several hours may travel some distance in order to ap-

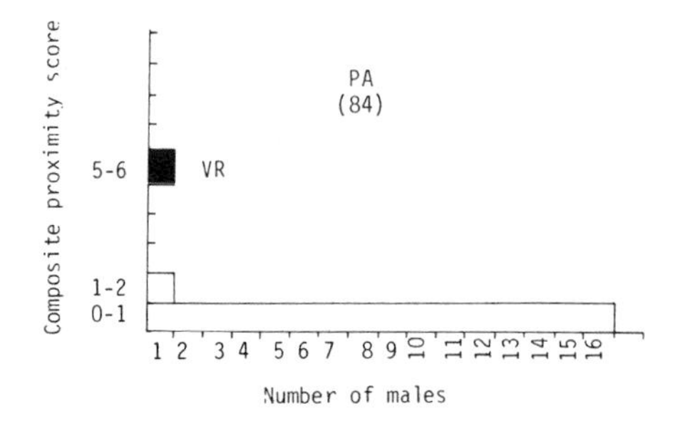

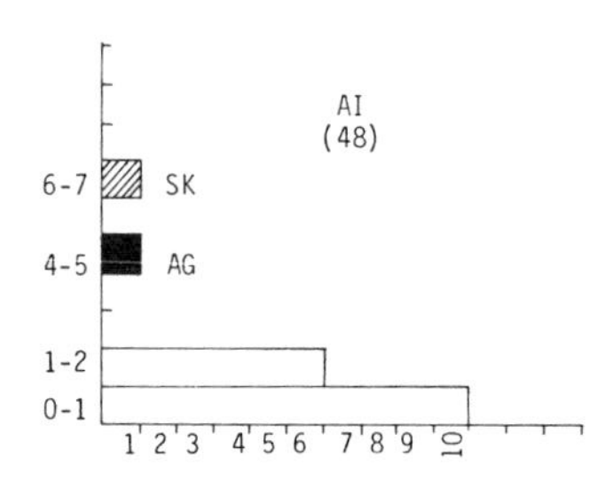

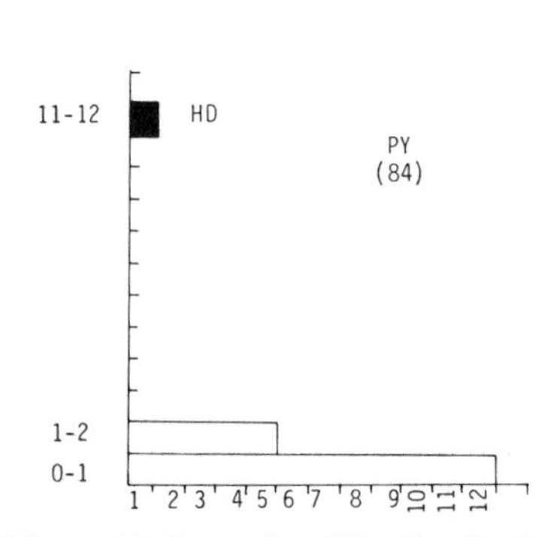

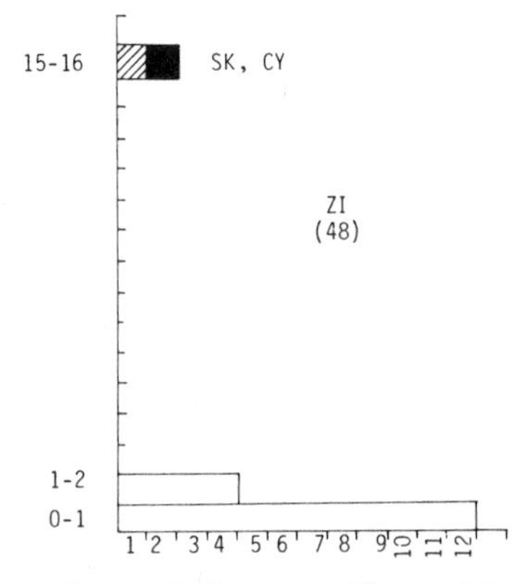

Figure 4.3. Examples of the distribution of male composite proximity scores (*C* scores) for four anestrous females. Each graph shows, for a particular female, the number of males (x axis) whose *C* score fell within a given interval (y axis) (see text and Appendixes VI and VII for method of determining *C* scores). In each graph, the initials of any male who had a particularly high *C* score are indicated to the right of the bar representing his score, and the female's primary and secondary grooming partners are indicated with a solid and a cross-hatched bar, respectively. The number of 1/2 hour focal female samples from which the *C* scores were derived is indicated in parentheses under the female's initials.

proach and either groom or solicit grooming from that partner. Thus it seems likely that grooming and proximity can be viewed as independent measures of affinity.

CRITERIA FOR FRIENDSHIP

The striking similarities between the ad lib grooming scores and composite proximity scores provided the basis for a definition of friendship. I began by identifying for each female a male or males who ranked either one or two on both the ad lib grooming and *C* scores. For the moment, I will refer to these males as "Affiliates." Based on these criteria, all but two of the 31 anestrous females had at least one male Affiliate (see Table 4.3). Without exception, this method isolated a male/female pair that I considered, subjectively, to be friends. However, there remained a number of pairs whom I thought of as friends but who were not identified by these criteria. I created a second category, "Associates," that included these males and then looked more closely at their grooming and *C* scores (Table 4.3). I found that every Associate scored high (rank four or above) on at least one of the two measures (grooming or proximity), and many scored high on the other as well. These results indicated that the objective characteristics of Associates were only slightly less outstanding than those of Affiliates.[3]

Did the distinction between Affiliates and Associates reflect real differences in baboon relationships, or was it simply an artifact of the criteria used to define friendship? The answer is probably a little of each. Any random sample of male–female dyads is likely to include some dyads in the process of forming a friendship and others in the process of ending one (see Chapter 9). Since friendships tend to begin and end gradually, some Associate dyads very well might have represented friendships in one of these transitional phases. A second possibility is that some Associate dyads were "friends" but not "best friends" (i.e., that some relationships were stronger than others). It would be surprising if this were not the case. A third possibility is that the differences between the scores of Affiliates and Associates simply reflected stochastic processes and random sampling biases.

Whether the distinction between Affiliates and Associates was real or a reflection of my methods, I suspected that the differences between dyads in these two categories were trivial for my purposes. I tested this hypothesis by comparing the scores of Affiliates and Associates on all of the measures described in Chapter 5. In no case were the scores of the

[3] The distinction between Affiliates and Associates may correspond approximately to the two types of special male–female relationships that Seyfarth identified in a small troop of chacma baboons: (1) persistent bonds characterized by very high frequencies of grooming and spatial proximity and (2) persistent bonds characterized by somewhat lower frequencies of these two behaviors (Seyfarth, 1978b). In the chacma baboon troop, however, these two types of relationships could be distinguished clearly, whereas in EC, grooming and proximity scores for Associates and Affiliates overlapped considerably.

Table 4.3. List of Male Friends (Affiliates and Associates)[a]

	Female	Male	Rank of male on ad lib grooming	Rank of male on C score	Affiliate	Associate
1.	AI	SK	2	1	+	
		AG	1	2	+	
2.	AT	HS	1.5	1	+	
		AG	3	2		+
		HC	4	3		+
3.	AU	AS	1	1	+	
4.	CB	HD	1	2	+	
		SK	3	1		+
5.	CC	CY	1	1	+	
6.	CG(1)[b]	IA	1	1	+	
	(1)	AC	>4	2		+
	CG(2)[b]	AS	2	1	+	
	(2)	SK	1	3		+
7.	CI	HC	1	1	+	
		VR	2	2	+	
8.	DD	HD	1	1	+	
		CY	2.5	2		+
9.	DL	HM	1	1	+	
		PX	2	2	+	
10.	DP	PX	1	1	+	
		HD	2	2	+	
11.	EU	VR	1	2	+	
		HC	2	1	+	
		AG	>4	3		+
12.	HH	BZ	2	1	+	
		SK	1	>4		+
13.	IO	PL	2	2	+	
		AS	1	4		+
		BZ	4	1		+
		AC	3	3		+
14.	IS	VR	1	2	+	
		HC	2	1	+	
15.	JO	HC	1	1	+	
16.	JU	SK	1	1	+	

Table 4.3. (Continued)

	Female	Male	Rank of male on ad lib grooming	Rank of male on C score	Affiliate	Associate
17.	LE(1)[b]	AC	1	2	+	
	(1)	AG	2	1	+	
	(1)	BZ	>4	3		+
	LE(2)[b]	AC	1	1	+	
	(2)	AS	2	3		+
	(2)	PL	3	2		+
18.	LI	HM	1	1	+	
		HD	2	3		+
		VR	3	2		+
19.	LU	VR	1	1	+	
20.	ML	PX	1	2	+	
		AS	2.5	1		+
		BZ	4	3		+
21.	MM	AC	1	1	+	
		AO	2	2	+	
22.	PA	VR	1	1	+	
23.	PH	AO	1	1	+	
		BZ	>4	2		+
24.	PO	CY	1	1	+	
		SK	2	2	+	
25.	PY	HD	1	1	+	
26.	OL	AC	1.5	2	+	
		AS	1.5	1	+	
27.	RH	BZ	1	1	+	
28.	XA	SK	Rarely groomed	1		+
29.	ZD	AO	1	1	+	
		HC	2	2	+	
30.	ZI	CY	1	2	+	
		SK	2	1	+	
31.	SO	None	Rarely groomed	No male with high score	—	—
	Total				43	20

[a]Two of the 33 females used in the analysis of focal animal samples (PS and ZN) are not included because they were cycling during most of the study period.

[b]Data for CG and LE are divided into two periods: (1) before and (2) after a P grooming partner disappeared from the troop.

two groups significantly different, and so I combined them into one category, "Friends." There were 62 Friend dyads, representing 12% of all possible male–female dyads.[4]

In the remainder of this chapter and all subsequent chapters, several conventions have been adopted to simplify discussion. First, all males who were not classified as Friends of a given female are referred to as Non-Friends. Second, these two terms, Friends and Non-Friends, are capitalized throughout in order to remind the reader that they have been formally defined. Third, whenever I refer to the identities of the members of a pair, I have listed the female's initials first (e.g., AT–HS).

CHARACTERISTICS OF FRIENDS

We have seen that the EC baboons really did have Friends, but who were they? Below, I consider which types of individuals were likely to be linked by friendship and whether some individuals were more popular as Friends than others.

Kinship

Four of the 62 Friend dyads listed in Table 4.3 (AT–HS, DP–PX, LI–HM, MM–AO) involved presumed mother–son relationships. In each case, the male was thought to have been born in EC, and the female was old enough to be his mother. Interactions between members of these pairs were particularly relaxed and stable. In each case, the male had a close relationship with all of the female's immature offspring (both known and putative; see Appendix VIII for method of assigning putative offspring), including unusually high frequencies of proximity, carrying, grooming, playing, and defense of the offspring by the male. Finally, in none of the pairs did the male exhibit any sexual interest in the female when she was in estrus. These same traits were characteristic of the two *known* mother–son relationships involving subadult males observed in EC in 1983. In some of the analyses that follow, I have commented separately on the results for the four putative mother–son dyads, and in other cases (particularly those involving sexual behavior) these pairs have been excluded from analysis.

[4] Manzolillo (1982) reported that, in the adjacent PHG troop, some male–female special relationships lasted for only a few months. Brief special relationships also existed in EC, although they were not so common as they were in PHG. Such brief relationships are not included in the analyses that follow; all Friend dyads showed high frequencies of grooming and spatial proximity throughout most or all of the study period. See Chapter 9 for a discussion of factors affecting the duration of special male–female relationships.

Excluding these four pairs, nearly all friendships involved either males that had transferred to EC from other troops or natal males paired with females too young to be their mothers. Friendships with maternal sisters can be ruled out for the four natal males whose mothers were identified, since none had a mature sister. It is possible that some natal males had friendships with paternal sisters, but there is no way to test this possibility. However, since 73% of the friendships involved nonnatal males, most Friends were not close relatives.

Number of Friends

Of the 31 females sampled during pregnancy/lactation, all but one, SO, had at least one male Friend.[5] The mean, median, and modal number of Friends was two. This is consistent with Altmann's (1980) report that most of the mothers in her troop of yellow baboons had special bonds with two males. Age had no effect on the number of Friends a female had, and females in the bottom one-half of the dominance hierarchy had only slightly fewer Friends (mean = 1.7) than did females in the top one-half of the hierarchy (mean = 2.0).

Although females varied little in the number of Friends they had, this was not the case for males. Adult, long-term residents had the greatest number of female Friends (mean = 5.2; median = 6); young adult, short-term residents had slightly fewer Friends (mean = 5.2; median = 5); young adult and subadult natal residents still fewer (mean = 3.5; median = 2.5); and young adult newcomers had none at all (these categories of males are based on Table 2.4). All adult, long-term residents had roughly the same number of female Friends, whereas young adult, short-term residents showed great variation in the number of Friends they had. For example, one short-term resident, SK, had eight Friends, two more than any other male, but another short-term resident, IA, had only one Friend. This result can be interpreted in two ways, which are not mutually exclusive. First, it may take some immigrant males much more longer than others to form friendships with several females. Second, it is possible that males who are successful in making Friends are more likely to remain in a troop. Chapter 9 provides support for both interpretations. (See Appendix IX for a list of each male's Friends and the age classes and dominance ranks of those females.)

[5] There was no obvious reason why SO did not have any male Friends. AN, the old menopausal female who disappeared from the troop for weeks at a time, did not have a male Friend either. All of the adolescent females not included as focal animals had a special relationship with at least one male, based on grooming data.

Types of Friends

Female age and dominance rank were significantly related to the age/residence status of male Friends [*11*, *12*]. The older the females, the greater the proportion of friendships they had with the older, long-term resident males; the younger the females, the greater the proportion of friendships they had with the younger, natal males (see Table 4.4). In other words, baboons tended to have friendships with partners who were approximately their own age.

Table 4.5 shows that high-ranking females, like older females, also had more friendships with the older, long-term resident males, and low-ranking females, like younger females, had more friendships with the younger, natal males. Since age and dominance were not correlated among these females, these results represent two independent trends.

The relationship between female characteristics and the proportion of friendships involving short-term residents (i.e., young adult, recently transferred males) is less straightforward. The youngest and oldest females had proportionately fewer relationships with short-term residents than did females of the two intermediate age classes (Table 4.4). High-ranking females also had slightly fewer friendships with short-term residents than did low-ranking females (Table 4.5). Both older females (i.e., all females except primipares) and high-ranking females, however, had proportionately more friendships with short-term residents than with natal males of equivalent age, a finding consistent with Packer's conclusion that in general immigrant males are more attractive to females than are natal males (Packer, 1979a,b).

The relationship between female rank and male age/residence status can also be investigated from the male point of view by comparing the numerical dominance ranks of the female Friends of different categories of males. The results show that the dominance ranks of female Friends were not significantly different between long- and short-term residents or between either group and *adult* natal males, but the female Friends of *subadult* natal males were significantly lower ranking than those of both long- and short-term residents [*13*].

There is an interesting pattern in the dominance ranks of females affiliated with the same males (see Appendix IX). For some males, especially HC, AG, VR, SK, and AC, there is a tendency for several of the female affiliates to rank near one another. Female baboons usually rank just below their mothers, and so females who have similar ranks tend to be closely related (Hausfater *et al.*, 1982). Since close relatives tend to associate with one another (Smuts, in preparation), males who make Friends with closely related females might find it easier to maintain proximity to several females than males who form bonds with

Table 4.4. Percentage of Male Friends from Different Age/Residence Categories for Females of Different Ages[a]

Female age class	Young adult and subadult natal males (N=6)[c]	Young adult short-term resident males (N=4)	Adult long-term resident males (N=5)
1. Primiparous (N=9)[b]	66.7	11.1	22.2
2. Young adult (N=22)	27.3	40.9	31.8
3. Middle-aged (N=25)	4.6	36.4	59.1
4. Old (N=6)	0	20.0	80.0

[a]Since only older females could have subadult or adult sons as Friends, putative mother–son dyads were eliminated from analysis.
[b]N=number of Friend dyads for female age class.
[c]N=number of males in age/residence category.

unrelated females. Similarly, closely related females might prefer to make Friends with the same male(s).

SUMMARY AND DISCUSSION

Measures of grooming and proximity were used to differentiate male–female pairs with particularly strong bonds from all other male–female dyads. Grooming between anestrous females and adult and subadult males was restricted mostly to one or two particular partners. This tendency was also apparent among cycling females not in consortship, but it was not as strong as it was among anestrous females. Most anestrous females were also found in proximity to one, two, or occasionally three adult males considerably more often than they were found near any other males. These were usually the same individuals the female groomed with most often. The striking overlap between male scores on grooming and proximity provided a basis for the definition of friendship. Friends included males who scored very high (one or two) on both measures (Affiliates) and also males whose scores were slightly lower but who were considered, subjectively, to have a strong bond with a particular female (Associates); other differences between these two groups were insignificant.

There was little variation among females in the number of Friends they had, but both older and higher-ranking females tended to have more friendships with older, long-term resident males, while young and lower-ranking females tended to have more friendships with younger, natal males. Adult, long-term residents—the males who were most often the Friends of older and higher-ranking females—were also the ones with the greatest number of female Friends. Natal males and

Table 4.5. Percentage of Male Friends from Different Age/Residence Categories for High- and Low-Ranking Females

Female dominance	Young adult and subadult natal males (N=6)[b]	Young adult short-term resident males (N=4)	Adult long-term resident males (N=5)
Top one-half of hierarchy (N=34)[a]	14.7	26.5	58.8
Bottom one-half of hierarchy (N=28)	42.9	35.7	21.4

[a]N=number of Friend dyads for high- and low-ranking females.
[b]N=number of males in age/residence category.

most short-term residents had somewhat fewer Friends, and newcomers had none at all.

Friends were members of a very select group: They accounted for only 12% of all possible pairs of anestrous females and males. In Chapter 5, interactions among Friends are compared to those of other male–female dyads in order to see just how special Friends really were.

NOTES ON STATISTICS

[*1*] The probability that an anestrous female would groom with a favorite grooming partner as often as observed (P_m) was calculated as follows. Under the null hypothesis, I assumed that for each grooming episode or "trial" the probability that a female would groom with male "m" (G_m) was equal to the proportion of all grooming episodes between males and anestrous females that involved male m. Using the formula for the binomial expansion,

$$P_m = \sum_{i=x}^{n} \binom{n}{i} (G_m)^i \, (1-G_m)^{n-i}$$

where n = the female's total number of grooming episodes with any male, x = the observed number of grooming episodes with male m, and G_m is defined as above. P_m values for different females are shown in Table 4.1.

[*2*] The probability that a male would groom with a favorite anestrous female partner as often as observed (P_f) was calculated using the method described above, reversing the roles of males and females. P_f values for different males are shown in Table 4.2.

[*3*] Comparison of number of females whose proportion of grooming with the P male was smaller when she was anestrus than when she was cycling, versus the reverse. Sign test: $x=4$, $N=22$, $p < .001$.

[*4*] Comparison of the cumulative frequency distribution of the amount of grooming accounted for by different numbers of males when females were cycling and when they were anestrus. Kolmogorov Smirnoff two-sample test: $d=.21$, $n_1=262$, $n_2=816$, $p < .001$.

[*5*] Comparison of the cumulative frequency distribution of the amount of grooming accounted for by different numbers of cycling females and anestrous females. Kolmogorov Smirnoff two-sample test: $d=.13$, $n_1=262$, $n_2=816$, $p < .005$.

[*6*] The probability that a cycling female would groom with a favorite male partner as often as observed was calculated using the method described in [*1*] above.

[*7*] The probability that a male would groom with a favorite cycling female partner as often as observed was calculated using the method described in [*1*] above.

[*8*] Correlation between female dominance rank and proportion of grooming done with P male (anestrous females). Spearman rank correlation coefficient: $r_s=.06$, $t=.34$, $d.f.=32$, $N=34$, n.s.

[*9*] Comparison of proportion of grooming done with P male for females of different ages (anestrous females). Mann–Whitney U test:

age class 1 versus 2: $U=39$, $n_1=8$, $n_2=10$, n.s.
age class 1 versus 3: $U=42$, $n_1=8$, $n_2=13$, n.s.
age class 1 versus 4: $U=7$, $n_1=8$, $n_2=3$, n.s.
age class 2 versus 3: $U=52$, $n_1=10$, $n_2=13$, n.s.
age class 2 versus 4: $U=24$, $n_1=10$, $n_2=3$, n.s.
age class 3 versus 4: $U=15$, $n_1=13$, $n_2=3$, n.s.

[*10*] Comparison of proportion of grooming with P partner for females who had a secondary partner versus those who did not (anestrous females). Mann–Whitney U test: $U=31$, $n_1=17$, $n_2=16$, $p < .001$ (one-tailed).

[*11*] Comparison of observed versus expected frequencies of male Friends belonging to three different age/residence classes (adult long-term residents; young adult short-term residents; young adult/subadult natal residents) for females of different age classes. $\chi^2=18.65$, $d.f.=6$, $N=58$, $p < .01$.

[*12*] Comparison of observed versus expected frequencies of male Friends belonging to three different age/residence classes (as above) for high- versus low-ranking females. $\chi^2=9.94$, $d.f.=2$, $N=62$, $p < .01$.

[*13*] Comparison of dominance ranks of female Friends of (*a*) subadult natal males versus adult long-term resident males and (*b*) subadult natal males versus young adult short-term resident males. Mann–Whitney U test:

(*a*) $U = 70.5$, $z = 2.6$, $n_1 = 12$, $n_2 = 26$, $p < .01$
(*b*) $U = 54.5$, $z = 2.4$, $n_1 = 12$, $n_2 = 19$, $p < .02$

5 WHAT MADE FRIENDS SPECIAL

Zena (right) and her friend Hector still asleep on the sleeping cliffs at dawn, in the same spot where the author left them the night before. Such physical intimacy is rare among most male–female pairs, but common among Friends.

INTRODUCTION

What made Friends special was, most of all, the unusual quality of their interactions. Female baboons, in general, are wary of males. This is understandable: Males sometimes use their larger size and formidable canines to intimidate and bully smaller troop members. Females, however, were apparently drawn to their male Friends, and they seemed surprisingly relaxed around these hulking companions. The males, too, seemed to undergo a subtle transformation when interacting with female Friends. They appeared less tense, more affectionate, and more sensitive to the behavior of their partners.

The goal of this chapter is to provide quantitative descriptions of some qualities of male–female relationships and to compare these descriptions for Friends and Non-Friends. Three types of behaviors are considered: (1) groomer/groomee roles; (2) movements that determined how much time a male and female spent in proximity; and (3) the types of interactions that occurred between males and females once they were in close proximity (less than 1 m apart). In my investigation of these behaviors, I will frequently draw on the work of Robert Hinde and his colleagues. Hinde, a professor of animal behavior at Cambridge University, England, has explored in detail the special methodological problems involved in providing objective descriptions of the qualities of social relationships (e.g., Hinde, 1977).

GROOMER/GROOMEE ROLES

One aspect of a relationship or interaction that Hinde has emphasized is the degree of reciprocity versus complementarity of roles: "A reciprocal interaction is one in which the participants show similar behaviour, either simultaneously or alternately, whereas in a complementary interaction the behaviour of one differs from, but complements, that of the others" (Hinde, 1976, p. 7). Hinde gives some examples: Play, in which individuals alternate roles of chaser and chased, is reciprocal; copulation, in which one individual always mounts and the other is always mounted, is complementary (Hinde, 1976, p. 7). A series of grooming interactions can be evaluated in similar terms: Do two individuals frequently alternate groomer/groomee roles, or does one individual consistently adopt one role and the other the other role? This chapter asks this question about grooming interactions between EC males and females. It also considers whether the frequency of grooming or female reproductive condition affects the degree of reciprocity/complementarity seen in these interactions.

For each ad lib grooming record, the groomer role was assigned to the individual grooming the other when the pair was first observed. I then determined for each anestrous female the proportion of grooming episodes with favorite grooming partners in which she groomed the male (recall that favorite grooming partners included only males with whom a female groomed on at least 20% of her grooming episodes with males). Of the 55 pairs, in only 4 was the male the groomer more often than he was the groomee [*1*]. When values for all females were averaged, females groomed males in 87% of the grooming episodes.

These values can be compared with values for groomer/groomee roles among infrequent grooming partners (i.e., males with whom the female groomed for less than 20% of all grooming episodes with males). There were 23 anestrous females who groomed with infrequent partners at least twice. In 20 of the 23, the female groomed the males more often than they groomed her [*2*]. When values for all females were averaged, females groomed males in 76% of the grooming episodes.

Results were similar for cycling females outside of sexual consortships. In 35 of 36 pairs, cycling females groomed their favorite male partners more often than the males groomed them [*3*], and when values were averaged over all females, females groomed males in 79% of the episodes. For 13 of 15 pairs of cycling females grooming with infrequent partners, the female groomed the male more than he groomed her [*4*], and on average the female groomed the male in 83% of the episodes.

These findings suggest two conclusions. First, it appears that male Friends (i.e., frequent grooming partners) were slightly less likely to adopt the groomer roles than were Non-Friends when the females were anestrus. These differences were small, however, and the more striking result is that no matter what her reproductive condition (excluding the consortship situation) and no matter who her partner, the female was far more likely to groom the male than he was to groom her.

These results are consistent with those of other studies. Females have been found to groom males more than the males groom them among chacma baboons (Hall, 1962; Saayman, 1971b), yellow baboons (Hausfater, 1975), rhesus monkeys (Kaufman, 1965, 1967; Lindburg, 1973), Japanese macaques (Oki and Maeda, 1973), bonnet macaques (Sugiyama, 1971), crab-eating macaques (Angst, 1975), mangabeys (Chalmers, 1968), and Sykes monkeys (Rowell, 1974). However, several authors have reported that males groom females more around the middle of the female's estrous cycle when ovulation is most likely to occur (baboons: Hall, 1962; Saayman, 1971a; Hausfater, 1975; Rowell, 1974; macaques: Michael *et al.*, 1966; chimpanzees: Goodall, 1968).

Seyfarth (1978b) reported that in two male–female pairs of chacma baboons, groomer/groomee roles reversed during sexual consortship. In these pairs, the proportion of male grooming was very low during pregnancy and lactation and comparable to the values reported above for EC. When consorting, however, male Rocky groomed the female on 100% of their grooming bouts, and male Pierre groomed the female on 80% of their grooming bouts. Similarly, Rasmussen (1980) found that males grooming females accounted for about 75% of all grooming bouts when cycling females were fully swollen, but almost no grooming bouts when cycling females were inflating or deflating (based on Rasmussen [1980; Figure 5.4, p. 5:10]). So, much as we might have expected, the nature of male–female relationships changes, even among Friends, when the female becomes sexually receptive to the male.

FEMALE AND MALE ROLES IN MAINTAINING PROXIMITY

In the previous chapter, I showed that most of the anestrous females in the troop spent a disproportionate amount of time in proximity to one, or a few, particular males. However, as Hinde has argued persuasively, this information alone does not tell us much about relationships (e.g., Hinde, 1977). Take, for example, a high school couple who spends considerable time together. This might occur because the boy is an unusually possessive person who stays near the girl to discourage the attention of other boys; because she is shy around boys in general and avoids all boys except her friend with whom she feels secure; or because they like each other and seek one another's company. These three reasons for their close proximity reflect three very different types of relationships, yet each might result in similar amounts of time in proximity.

This example shows why it is important to relate information about time spent in proximity to information about how proximity is achieved and maintained. When this is done, it is possible to answer four questions about spatial proximity between two individuals, *A* and *B*:

1. Which individual, *A* or *B*, is primarily responsible for the fact that these two spend considerable time together?
2. Suppose the answer to the first question is *A*. We can then ask whether *B* also contributes to proximity between the two, plays a neutral role, or avoids *A*, reducing the amount of time they spend together.
3. Are differences in the amount of time *A* and *B* spend together

compared to the amount of time they spend near others due primarily to *A*'s behavior or to *B*'s? Note that this is a different question from question 1 above, which was concerned only with *A* and *B* and not with a comparison of their behavior with that of others.
4. Suppose the answer to question 3 is *A*'s behavior. We can then ask whether differences in proximity are due primarily to *A*'s preferences or aversions, or both (e.g., *A* could seek proximity to both *B* and others, but seek proximity to *B* more, or *A* could avoid both *B* and others, but avoid others more).

The first two questions concern only the relationship between *A* and *B*, while the latter two questions allow a comparison of *A* and *B*'s relationship with other relationships. These questions are used, below, to provide a framework for describing the behavioral dynamics underlying (*a*) spatial proximity among Friends and (*b*) differences in spatial proximity between Friends and Non-Friends.

All of the analyses that follow depend on two basic measures: the frequency with which one individual moved toward another (approaches) and the frequency with which that same individual moved away from another (leaves). These measures were derived from focal samples of adult females. Whenever proximity between the female and a male changed from one of four distance categories to another, I noted the individual whose movement caused the change (the four distance categories were: one: 0–1 m; two: 1–2 m; three: 2–5 m; four: > 5 m). In general, movements that resulted in an increase in proximity from any distance category to a smaller category were scored as approaches, and movements that resulted in a decrease in proximity from any distance category to a larger one were scored as leaves. Some restrictions were applied to this procedure to ensure unbiased samples (see Appendix X for details).

Movements into and out of distance category one ("arm's reach") were frequently associated with overt social interactions and seemed qualitatively different from shifts in proximity at greater distances. For this reason, such movements, termed "close" *approaches* and *leaves*, were analyzed separately from movements over greater distances. I begin with these latter movements, returning to examine close approaches and leaves in a later section.

Responsibility for Maintaining Proximity Over 1–5 Meters

Hinde and his colleagues have shown that the most useful general measure for analyzing roles in maintaining (or breaking) proximity is the percentage of all approaches between two individuals, x and y, that were due to x minus the percentage of all leaves due to x ($\%A_x - \%L_x$). When the value of this index is positive, it means that x was primarily responsible for maintaining proximity between the pair; when the index is negative, it means that y was primarily responsible for proximity (e.g., see Hinde, 1977). Since the percentage of approaches (or leaves) due to x and the percentage due to y must always sum to 100, the $\%A - \%L$ index can be calculated from scores of either x or y. As a convention, I have always used the movements of the female to determine this index for each dyad.

Table 5.1 shows for each anestrous female her $\%A_f - \%L_f$ index for each Friend and for Non-Friend males. Since the frequency of approaches and leaves involving Non-Friend males was small, for each female data for all Non-Friend males were combined. The term *Non-Friend dyads* below always refers to these summed scores.

1. *Which individual, the female or the male, was primarily responsible for spatial proximity between Friends?* In most (85%) of the Friend dyads, the $\%A - \%L$ index was positive, indicating that the female was primarily responsible for maintaining proximity between Friends [*5*].

2. *Did males also contribute to proximity between Friends?* To answer this question we need to look at the *absolute* number of approaches males made to Friends compared to the number made to Non-Friends, per unit of female observation time (Hinde, 1977, pp. 5–6). For 69% of the Friend dyads, the female was approached by her Friend more often than by any other male (excluding other Friends) [*6*], and for 96% of the Friend dyads, the female was approached by her Friend more often than she was approached on average by Non-Friend males [7]. These results provide clear evidence that, although males usually were less responsible for proximity between Friends than were females, they nevertheless contributed to this proximity.

3. *Which sex was primarily responsible for differences in the amount of time Friends and Non-Friends spent together?* Hinde (1977) has shown that when differences between individuals in time spent in proximity and differences in *A*'s role in maintaining proximity to these different individuals lie in the same direction (e.g., both decrease), then we can conclude that the differences in time spent in proximity are primarily due to *A*'s behavior. In this case, if the female's role in maintaining proximity to Non-Friend males was less than her role in maintaining proximity to Friends, we can conclude that females were primarily responsible for the fact that Non-Friends spent less time together than did Friends. The results support this conclusion: in 83%

Table 5.1. Responsibility for Maintaining Proximity between Adult Males and Females Over Distances of 1–5 Meters[a]

Female	Male	$\%A_f - \%L_f$	N
AI	SK	39.3	34
	AG	29.9	34
	NF	−2.9	70
AT	NF	−13.4	30
AU	AS	3.2	48
	NF	1.6	65
CB	SK	0	30
	NF	−28.2	88
CC	CY	44.8	25
	NF	−11.3	52
CG	IA	−13.6	25
	SK	47.3	40
	AC	27.2	88
	NF	−0.1	140
CI	NF	−26.7	46
DD	HD	41.7	48
	NF	−15.3	68
DL	HM	41.4	75
	NF	−24.5	91
DP	PX	25.2	38
	HD	17.2	56
	NF	−3.3	124
EU	VR	31.6	68
	HC	30.0	55
	AG	41.0	50
	NF	−13.4	77
IO	BZ	42.7	46
	AC	28.2	22
	NF	−6.7	43
IS	HC	31.5	33
	VR	41.7	22
	NF	−12.4	29
JO	HC	45.0	44
	NF	−7.1	60
JU	NF	−12.1	55
LE	AC	4.9	30
	BZ	6.7	50
	NF	10.2	173
LU	VR	58.7	39
	NF	−3.3	53
ML	BZ	57.0	28
	NF	0	32
MM	AC	29.0	51
	AO	17.2	51
	NF	−20.5	100
OL	AC	11.2	34
	AS	−25.0	24
	NF	18.5	62
PA	VR	80.9	42
	NF	−54.2	35
PH	AO	−2.1	98
	BZ	25.6	62
	NF	−5.7	140
PO	CY	27.3	295
	SK	3.4	41
	NF	2.3	174
PY	HD	33.3	114
	NF	−14.0	133
RH	BZ	39.1	178
	NF	10.1	54
SO	NF	8.8	137
XA	SK	34.5	132
	NF	−23.1	65
ZD	AO	−42.0	68
	HC	−45.7	37
	NF	−18.8	165
ZI	SK	12.6	123
	CY	40.8	48
	NF	−23.1	62

[a]The percentage of approaches made by the female minus the percentage of leaves made by the female ($\%A_f - \%L_f$) is shown for each female, for each Friend dyad, and for all Non-Friends combined. N = total number of approaches and leaves. Following Hinde and Proctor (1977, p. 305), dyads with fewer than 20 approaches + leaves were eliminated; thus for some females only data for Non-Friend males are shown.

Table 5.2. Signs of $\%A_f - \%L_f$ Index for Friends and Non-Friends and Their Interpretation for Friend Dyads in Which the Female Was Primarily Responsible for the Greater Proximity between Friends Compared with Non-Friends

$\%A_f - \%L_f$ Friends	$\%A_f - \%L_f$ Non-Friends	Number of Friend dyads	Interpretation
+[a]	–	24	Female sought Friend and avoided Non-Friends
+	+	4	Female sought both Friend and Non-Friends but sought Friend more
–[b]	–	1	Female avoided both Friend and Non-Friends but avoided Non-Friends more
+	0	1	Female sought Friend and neither sought nor avoided Non-Friends
0[c]	–	1	Female neither sought nor avoided Friend but avoided Non-Friends
Total		31	

[a] $\%A_f - \%L_f > 0$
[b] $\%A_f - \%L_f < 0$
[c] $\%A_f - \%L_f = 0$

of the Friend dyads, the $\%A - \%L$ value for Non-Friends was less than the value for the Friend (see Table 5.1) [*8*].

4. *Among Friend dyads in which the female was primarily responsible for the fact that Friends spent more time together than Non-Friends, was this due to female preferences, aversions, or both?* To answer this question, we need to compare the sign of $\%A - \%L$ for Friends and Non-Friends for the 31 Friend dyads in which this index was greater for Friends than for Non-Friends[1] (see Table 5.2) (Hinde, 1977, pp. 3–5). For 77% of these dyads, the results indicate that Friends spent more time together than Non-Friends because the female *sought* proximity to her Friend and *avoided* proximity to Non-Friends.

[1] A similar analysis could, in principle, be done for the seven pairs in which the male was primarily responsible for the greater proximity of the Friend dyad, but the data were not available due to the absence of focal samples on males.

Table 5.3. Responsibility for Maintaining Close Proximity between Adult Males and Females (Distances of 0–1 Meter)[a]

Female	Male	$\%A_f - \%L_f$	N	Female	Male	$\%A_f - \%L_f$	N
AI	SK	−36.3	22	JU	NF	−58.8	34
	NF	−60.0	20	LE	NF	−59.1	88
AT	NF	−53.8	26	LU	VR	36.4	22
AU	NF	−46.1	26		NF	−50.0	32
CB	NF	−56.3	32	MM	AC	−30.0	20
CC	NF	−55.6	26		AO	−10.0	20
CG	AC	−63.6	22		NF	−83.3	24
	NF	−30.0	40	PH	AO	−50.0	36
DD	HD	−40.0	20		NF	−75.0	40
	NF	−42.9	28	PO	CY	1.4	146
DL	HM	14.3	28		SK	−13.3	30
	PX	6.6	30		NF	−28.5	28
	NF	−53.0	34	PY	HD	13.7	44
DP	PX	−8.7	46		NF	−28.6	42
	NF	0	24	RH	BZ	35.4	96
EU	VR	50.0	20		NF	−20.0	20
	HC	0	38	SO	NF	−53.3	30
	AG	−7.7	26	XA	SK	17.0	60
IO	BZ	−16.6	24	ZD	AO	−70.4	54
	AC	−12.5	28		NF	−48.4	100
IS	HC	−10.0	20	ZI	SK	0	50
JO	HC	−15.8	38		CY	9.1	22

[a] N=total number of approaches and leaves. Dyads with fewer than 20 approaches and leaves were eliminated. Thus, for some females only data for Non-Friend males are given, and for others only data for Friends are given. Four of the females listed in Table 5.1 are not shown here because of inadequate sample sizes for both Friends and Non-Friends.

Taken together, these results suggest that, in general, anestrous females preferred to remain within 1–5 m of male Friends but preferred to be at distances greater than 5 m from all other males. Do the same results hold for close proximity?

Responsibility for Maintaining Close Proximity Over 0–1 Meter

I determined the $\%A - \%L$ index for Friends and for Non-Friends as in the previous section. Again, for each female, scores for all Non-Friend males were combined. The results are shown in Table 5.3.

1. *Which individual, the female or the male, was primarily responsible for close proximity among Friends?* In over one-half (62%) of the Friend dyads, the %*A* − %*L* index was *negative*, indicating that males were primarily responsible for maintaining proximity in the majority of cases, but this result was not statistically significant [*9*]. These data therefore do not allow rejection of the null hypothesis that there was no *overall* difference between male and female roles in maintaining close proximity, although within many dyads, one partner was clearly more responsible for maintaining close proximity than the other.

2. *Did both sexes contribute to proximity between Friends?* Although males were primarily responsible for maintaining close proximity in some Friend dyads and females in others, a look at the frequency of approaches by males and females shows that in nearly all Friend dyads, both partners contributed to close proximity. Females approached the Friend more often than they approached any Non-Friend in 74% of the Friend dyads [*10*], and females approached the Friend more often than they approached Non-Friends on average in 88% of the Friend dyads [*11*]. Similarly, males approached the Friend more often than they approached any Non-Friend in 81% of the Friend dyads [*12*], and males approached the Friend more often than they approached Non-Friends on average in 96% of the Friend dyads [*13*].

3. *Which sex was primarily responsible for differences in the amount of time Friends and Non-Friends spent in close proximity?* For 80% of the Friend dyads for which adequate data were available ($N = 15$), the %*A* − %*L* value for Non-Friends was less than the value for Friends (see Table 5.3) [*14*]. This means that females were primarily responsible for *differences* in close proximity between Friends and Non-Friends even though, more often than not, the females were not primarily responsible for close proximity between Friends. This suggests that females showed strong aversions to close proximity to Non-Friends. This hypothesis is tested below.

4. *Among Friend dyads in which the female was primarily responsible for the fact that Friends spent more time in close proximity than did Non-Friends, was this due to female preference, aversions, or both?* Comparison of the signs of the %*A* − %*L* index for Friends and Non-Friends indicated that one-half of the 12 Friend dyads spent more time in close proximity than did Non-Friends because the female *sought* close proximity to her Friend and *avoided* close proximity to Non-Friends. The other one-half of the Friend dyads spent more time in close proximity because the female *avoided* close proximity to *both* Friends and Non-Friends but avoided Non-Friends more. This tendency to avoid Non-Friend males was very strong: Among all females

who had at least 20 approaches and leaves involving Non-Friend males ($N = 19$), the $\%A - \%L$ value was negative for all but one, and the value for that female was zero. Thus, *no* anestrous females sought close proximity to Non-Friend males, and all but one actively avoided such proximity.

In summary, the movements of both females and males indicated consistent preferences for Friends relative to Non-Friends. Over distances of 1–5 m, females were primarily responsible for maintaining proximity to Friends, but in the majority of Friend dyads, males were primarily responsible for maintaining close proximity. These results suggest that although most females preferred to be near their male Friends, more than one-half of the females did not like being too close to them.

Even though they did not bear primary responsibility for the maintenance of close proximity to Friends, females were primarily responsible for *differences* in the amount of time Friends spent in close proximity compared with Non-Friend males; this was mainly due to strong aversions to close proximity to Non-Friend males. The ways in which these aversions affected the amount of time spent in close proximity is examined further in the next section.

Timing of Leave After a Close Approach

All females spent more time in close proximity (within 1 m) to Friends than to Non-Friends (see Appendix XI). This result could simply be a function of the greater number of close approaches by and to Friends, but it could also reflect differences in how long the male and female remained in proximity following an approach. In this section, the timing of the leave following a close approach is examined. The analysis includes only close approaches that occurred when the female was pregnant or lactating, where the corresponding leave was observed, and where neither the approach nor leave was mutual.

Length of Time Spent in Proximity After an Approach. Close approaches involving both Friends and Non-Friend males were usually followed by a leave within 1 minute, but this was more often true for Non-Friend males (88%) than for Friends (78%). When percentages for Non-Friend males and Friends were compared within females, the two classes of males were significantly different [*15*]. Thus, Friends remained in proximity for more than 1 minute more often than did Non-Friend dyads. In order to find out why this was the case, the timing of the leave was examined in more detail.

Close Approach Elicits an Avoid. An avoid was scored when one member of a male–female pair moved more than 1 m away from the other either upon the other's approach or immediately following a brief interaction (e.g., a present or a greeting). Avoids included "supplants"—interactions in which the approaching animal took over the departing animal's feeding or resting site. Male avoids of females were uncommon (males avoided only 1.5% of all female approaches), so only instances where the female avoided the male are discussed here.

For each female, the percentage of male approaches that elicited an avoid was determined for each Friend (41 dyads) and for all Non-Friend males combined (27 females). Mean percentages were then calculated for each type of male across all females. Females avoided Non-Friend males 50% of the time, more than twice as often as they avoided Friends (22% of the time). Since the percentage of avoids varied greatly among females, in order to test for significance each female's percentage of avoids for each Friend was compared with the same measure for Non-Friend males. In 32 of the 39 Friend dyads used in the comparison (females with fewer than five approaches for either the Friend or Non-Friend males were eliminated), the female avoided the approaches of Non-Friend males more often than she avoided the approaches of Friends [*16*]. This finding may hold true in other troops as well; among olive baboons at Gombe, females also avoided most frequently those males that associated with them the least (Packer and Pusey, 1979).

Approacher Leaves Immediately after Approach. An "immediate leave" was scored whenever the animal who approached and the animal who moved away first *were the same individual*, and the leave occurred within 5 seconds of the approach, whether or not a brief interaction preceded the leave. The proportion of all approaches by males and females that were not avoided but that were followed by an "immediate leave" was compared for Friends and Non-Friend males (see Appendixes XII and XIII). Most male approaches to Non-Friend females (67%) that the female did not avoid were followed by an "immediate leave," but this was the case for only a small proportion (12%) of male approaches to Friends. The differences between Friends and Non-Friends were even more striking for females: After approaching Non-Friend males, females moved away immediately 85% of the time, but this was the case for only 9% of the approaches they made to Friends. Differences between Friends and Non-Friends were significant for both sexes [*17, 18*].

Animals Remain in Proximity for More Than 1 Minute. I next determined the percentage of approaches not followed by an avoid or immediate leave that were followed by maintenance of proximity for more than 1 minute. The average score (28%) was exactly the same for approaches involving Friends and Non-Friend males, and the within-female comparison of scores did not show any significant difference between these two groups [*19*]. If an approach was not avoided or followed by an immediate leave, then Non-Friend dyads and Friends were equally likely to remain in proximity for more than 1 minute. Among Non-Friends, however, 84% of all approaches were followed by an avoid or immediate leave compared to only 22% of approaches among Friends.

Discussion

This section has shown that Friends remained in close proximity for longer than 1 minute more often than did Non-Friends because: (1) females were more likely to avoid the approach of Non-Friends than of Friends, and (2) both females and males were more likely to leave soon after their own approach when the approached animal was a Non-Friend than when s/he was a Friend. The fact that Friends spent more time in close proximity than Non-Friends is at least in part a reflection of these differences and not simply a result of more frequent approaches by and to Friends.

The results presented above also suggest that males and females approached Friends and Non-Friends for different reasons: It seems that Non-Friends usually approached to engage in a brief interaction with the other animal, whereas Friends often approached in order to spend time with their partner. While it is impossible to be sure of the motivation of an approaching animal, a closer look at the types of interactions that occurred following a close approach may shed further light on the nature of relationships.

INTERACTIONS FOLLOWING A CLOSE APPROACH: FRIENDS COMPARED WITH NON-FRIENDS

For each pregnant/lactating female, each approach by or to a male was scored for the presence of 19 types of interactions (see Table 5.4). These interactions were not mutually exclusive, and more than one could occur following an approach. Usually, however, approaches were followed by only one type of interaction.

For each female, the percentage of all approaches to Friends and Non-Friends that was followed by each type of interaction was determined. These values were then averaged over all females who had at

Table 5.4. Frequency of Different Interactions Following an Approach for Friends and Non-Friends[a]

Interaction	Friends	Non-Friend Males	N	No. of pluses	p
Feed together	38.8	24.8	28	21	**
Rest together	18.7	16.2	24	15	
Groom	14.5	4.5	25	21	**
Travel together	10.8	4.9	24	21	**
Male supplants female[b]	6.7	25.4	26	4	**
Male interacts with infant	4.8	2.7	20	13	
Female avoids male[b]	3.1	5.2	19	10	
Groom present—no groom	2.6	0.4	15	13	**
Female presents, male responds	1.7	8.1	18	4	
Third party interaction[c]	1.6	4.3	18	5	
Female submission[d]	1.1	7.0	19	2	**
Male inspects perineum[e]	0.8	3.2	11	3	
No interaction[f]	0.7	1.9	11	5	
Female approach to avoid other	0.7	0.0	5	5	
Nonsexual greet	0.5	0.3	3	2	
Drink together	0.3	0.6	5	3	
Male aggression	0.3	0.3	4	3	
Male mounts female	0.2	1.2	5	0	
Female presents, male ignores	0.2	4.0	10	0	**

[a]Columns two and three show the mean percentage for Friends and Non-Friends of all male–female close approaches that were followed by each type of interaction (percentages were found for each female first, and these values were then averaged). In order to determine whether the frequency of different interactions was significantly different for Friends versus Non-Friend males, the percentage of approaches followed by each type of interaction was compared *within* each female for Friends versus Non-Friends. When a female had zero scores for a particular type of interaction for both Friends and Non-Friends, she was eliminated from analysis for that interaction; this resulted in variable numbers of females for the within-female comparison (fourth column, N). The fifth column, No. of pluses, shows the number of females whose scores were higher for Friends than for Non-Friend males. The last column shows the significance level for each comparison; a blank indicates no significant difference, and ** indicates $p < .01$. See Chapter 5, [20], for statistical tests.

[b]Male supplants female was scored when the female avoided the male (see text) and the male took over her feeding or resting site. Female avoids male was scored when the female avoided the male and he did not take over her site.

[c]Third party interaction was scored when the approacher did not interact with the approachee but interacted with another baboon nearby.

[d]Female submission was scored when the female responded to the male's proximity with a submissive gesture (see Appendix III) in the absence of any aggression by the male.

[e]Male inspects perineum was scored when the male inspected the female's perineum in the absence of a female present.

[f]No interaction was scored when the approacher paused briefly and then moved away without interacting with the approached animal or any other nearby baboon.

least five approaches for both Friends combined and Non-Friend males combined ($N = 26$ females), and the results were compared for Friends and Non-Friends. Of the 19 interactions, 4 occurred at significantly higher frequencies when a Friend was involved: feeding together, travelling together, grooming, and groom present–no groom [*20*]. Three occurred at a significantly higher frequency when Non-Friend males were involved: female is supplanted by male, female presents[2] and male ignores her, and female submission [*20*]. These results, combined with the nonsignificant trends in interaction frequencies (see Table 5.4), indicate that Friends were more likely to engage in either routine maintenance activities (feed, rest, travel) or friendly nonsexual interactions (grooming, male interacts with infant), whereas Non-Friend dyads were more likely to engage in agonistic interactions (female supplanted by male, female submission) or sexual and/or appeasement interactions (female present, male inspect female perineum, male mount female).

The frequencies of appeasement/sexual interactions are particularly intriguing. Females presented to Non-Friend males about six times as often as to Friends (12% of approaches versus 2%). The higher frequency of presents to Non-Friend males could reflect the fact that these males were less familiar with the perineal condition of the female (as suggested by the fact that unsolicited inspections of the perineum were four times more common for Non-Friend dyads)—perhaps these females presented more often in order to give the males an opportunity to inspect them. However, males ignored the presents of Non-Friend females about one-third of the time compared with only about one-tenth of the less common presents of Friends.

There are two possible interpretations of these patterns. Either males are for some reason less interested in the perineums of Non-Friend females, or a male's response to a present by an anestrous female is primarily a social and not a sexual gesture—that is, he responds in order to communicate something to the female, rather than to gain information about her reproductive condition. What, then, is being communicated? My impression is that anestrous females (and often estrous females as well) used the present as a means to achieve proximity to males (or to respond to proximity of males) who made them nervous. It often seemed like the thing to do when a female did not know what else to do, when she was both drawn to a male and frightened of him at the same time. Female presents were often

[2] See Appendix III, Section III, E, for definitions of present and groom-present.

accompanied by signs of tension and ambivalence such as small hesitant movements toward and away from the male and nervous glances in his direction. If a female is in conflict about being close to a male, then a present may function as a way of responding to him that is less direct and assertive than a face-to-face encounter. It may also serve as a request for reassurance, for a friendly acknowledgment of her presence. If these speculations are correct, then the observed frequency of presents and responses could indicate: (*a*) that females were more ambivalent about the proximity of Non-Friend males; and (*b*) that males were more highly motivated to communicate "polite" acknowledgment of a female's presence when she was a Friend. A more detailed analysis of female presents is necessary to evaluate the suggestions presented here.

SUMMARY AND DISCUSSION

Grooming interactions among Friends were complementary rather than reciprocal: Males usually adopted one role (groomee) and females the other (groomer). In this regard, Friends did not differ from Non-Friends, who showed the same pattern of females grooming males—a pattern characteristic of many primate species. The main exception to this pattern involves females at the peak of estrus: Other studies indicate that sexually active couples switch roles, and the male becomes the primary groomer, the female the groomee (males also sometimes groomed females more in the early stages of a friendship; see Chapter 9). Thus, in terms of grooming, Friends differed from Non-Friends in the frequency with which they groomed together but not in the roles they adopted.

When we consider spatial proximity, a different pattern emerges. Friends differed from Non-Friends not only in the frequency with which they were found near one another but also in the roles males and females played in maintaining proximity. These differences applied both to close proximity and to proximity over distances of 1–5 m. Among Friends, both females and males contributed to maintaining proximity, although the female role was stronger over distances of 1–5 m and the male role was stronger in maintaining close proximity. Among Non-Friends, in general only males contributed to proximity. Females avoided Non-Friend males, especially when they approached to within 1 m, and female aversion to Non-Friend males was the single most important cause of reduced proximity among Non-Friend dyads compared with Friends.

When we consider the behavior of a couple following a close approach, Friends and Non-Friends again showed strongly divergent patterns. Non-Friends tended to separate quickly, sometimes after a tense interaction in which the female expressed fear, submission, or appeasement or, more often, because the female moved away before an interaction could take place. Friends, in contrast, were more likely to remain in close proximity, and while together they interacted in a relaxed, friendly manner or simply continued routine activities side by side.

The most striking result to emerge from this comparison of Friends and Non-Friends is the females' apparent desire to be near their Friends. In order to achieve this goal, females tailored their movements to those of their Friends. The desire to have male Friends nearby may also help to explain why females groomed male Friends so much more often than their Friends groomed them: Perhaps females provide males with the benefits of grooming in exchange for some advantage that females derive from male proximity.[3] What might this advantage be? More generally, what do females gain from Friendships with males, and what do males gain from Friendships with females? These questions are addressed in Chapters 6–9.

NOTES ON STATISTICS

[*1*] Comparison of number of pairs of anestrous females and favorite male grooming partners in which the male was the groomer more often than he was the groomee versus the number in which the reverse was true. Sign test: $x=4$, $z=6.20$, $N=55$, $p < .001$.

[*2*] Comparison of the number of pairs of anestrous females and infrequent male grooming partners in which the male was the groomer more often than he was the groomee versus the number in which the reverse was true. Sign test: $x=3$, $N=23$, $p < .001$.

[*3*] Comparison of the number of pairs of cycling females and favorite males in which the male was the groomer more often than he was the groomee versus the number in which the reverse was true. Sign test: $x=1$, $z=5.80$, $N=36$, $p < .001$.

[*4*] Comparison of the number of pairs of cycling females and infrequent male grooming partners in which the male was the groomer more often than he was the groomee versus the number in which the reverse was true. Sign test: $x=2$, $N=15$, $p < .01$.

[3] While this suggestion may sound anthropomorphic, there is increasing evidence that nonhuman primates do sometimes use grooming instrumentally to achieve social benefits (Seyfarth, 1976; Cheney, 1978; Reynolds, 1981; and especially Seyfarth and Cheney, 1984).

[*5*] Comparison of the number of Friend dyads in which $\%A_f - \%L_f$ was positive versus the number in which this index was negative (movements over distances of 1–5 m). Sign test: $x = 34$, $z = 4.48$, $N = 39$, $p < .001$.

[*6*] Comparison of the number of Friend dyads in which the female was approached by the male more often than she was approached by any Non-Friend male versus the number in which the reverse was true. Sign test: $x = 31$, $z = 2.39$, $N = 45$, $p < .01$.

[*7*] Comparison of the number of Friend dyads in which the female was approached by the Friend more than she was approached by the average Non-Friend male versus the number in which the reverse was true. Sign test: $x = 43$, $z = 6.26$, $N = 45$, $p < .001$.

[*8*] Comparison of the number of Friend dyads in which $\%A_f - \%L_f$ was greater than the same index for Non-Friend males versus the number in which the reverse was true. Sign test: $x = 33$, $z = 3.96$, $N = 40$, $p < .001$.

[*9*] Comparison of the number of Friend dyads in which $\%A_f - \%L_f$ was positive versus the number in which it was negative (close proximity). Sign test: $x = 9$, $N = 23$, n.s.

[*10*] Comparison of the number of Friend dyads in which the female approached the male more often than she approached any Non-Friend male versus the number in which the reverse was true (close proximity). Sign test: $x = 39$, $z = 3.30$, $N = 53$, $p < .001$.

[*11*] Comparison of the number of Friend dyads in which the female approached the male more often than she approached Non-Friend males on average versus the number in which the reverse was true (close proximity). Sign test: $x = 46$, $z = 5.4$, $N = 52$, $p < .001$.

[*12*] Comparison of the number of Friend dyads in which the male approached the female more often than he approached any Non-Friend female versus the number in which the reverse was true (close proximity). Sign test: $x = 43$, $z = 4.40$, $N = 53$, $p < .001$.

[*13*] Comparison of the number of Friend dyads in which the male approached the female more often than he approached Non-Friend females on average versus the number in which the reverse was true (close proximity). Sign test: $x = 50$, $z = 6.51$, $N = 52$, $p < .001$.

[*14*] Comparison of the number of Friend dyads in which $\%A_f - \%L_f$ was greater for Friends than the same index for Non-Friend males versus the number in which the reverse was true (close proximity). Sign test: $x = 12$, $N = 15$, $p < .05$.

[*15*] Comparison of the number of dyads in which a close approach involving a Friend was followed by a leave within 1 minute more often

than a close approach involving Non-Friend males versus the number of dyads in which the reverse was true. Sign test: $x = 20$, $N = 24$, $p < .002$.

[*16*] Comparison of the number of dyads in which the female avoided close approaches by Non-Friend males more often than she avoided approaches by Friends versus the number of dyads in which the reverse was true. Sign test: $x = 32$, $z = 4.06$, $N = 38$, $p < .001$.

[*17*] Comparison of the frequency of "immediate leaves" (see text for definition) by males for Friends versus Non-Friends (see Appendix XII for frequencies). $\chi^2 = 71.70$, $d.f. = 1$, $N = 225$, $p < .001$.

[*18*] Comparison of frequency of "immediate leaves" by females for Friends versus Non-Friends (see Appendix XIII for frequencies). $\chi^2 = 85.33$, $d.f. = 1$, $N = 172$, $p < .001$.

[*19*] Comparison of the number of dyads in which x was greater for Friends than for Non-Friends, where x = the number of close approaches that were not followed by either an avoid or an immediate leave that were followed by the male and female remaining in proximity for more than 1 minute versus the number in which the reverse was true. Sign test: $x = 13$, $N = 20$, n.s.

[*20*] Comparison of the number of females in which an interaction following a close approach occurred at a higher (or lower) frequency for Friends than for Non-Friends versus the number in which the reverse was true. Values are given only for interactions that showed significant differences (see Table 5.4). Sign test, interactions more frequent among Friends:

(*a*) Feed together: $x = 21$, $z = 2.46$, $N = 28$, $p < .02$
(*b*) Groom together: $x = 21$, $N = 25$, $p < .002$
(*c*) Travel together: $x = 21$, $N = 24$, $p < .002$
(*d*) Groom present-no groom: $x = 13$, $N = 15$, $p < .008$

Sign test, interactions more frequent among Non-Friends:

(*a*) Male supplants female: $x = 22$, $z = 3.33$, $N = 26$, $p < .001$
(*b*) Female submission: $x = 17$, $N = 19$, $p < .002$
(*c*) Female presents, male ignores her: $x = 10$, $N = 10$, $p < .002$

6 BENEFITS OF FRIENDSHIP TO THE FEMALE

A large subadult male, Pliny, holds the hand of Despoena, daughter of his Friend, Daphne. Males develop strong attachments to their Friends' infants, and they protect these small companions from aggression by other baboons.

INTRODUCTION

Why do female baboons form friendships with males? As social primates with friendships of our own, we may feel that the answer to this question is either obvious (how could such relationships not exist among intelligent long-lived animals who live in permanent social groups?) or outside the domain of science (they exist because some baboons just happen to like each other). For a psychologist interested in the development of relationships through time, the answer to this question might lie in the individual histories of the animals since birth—a worthwhile and potentially fascinating area for research, largely untapped in studies of animals. From the point of view of a biologist interested in the evolution of behavioral tendencies, the question can be rephrased as follows: How might having a friendship with a male increase the reproductive success of a female baboon?

Perhaps the best way to answer this question would be to compare directly the reproductive success of females who had Friends with that of females who did not. Such a comparison is not possible because there was only one female who did not have at least one male Friend. What I have done, instead, is to examine two ways in which Friends appeared to provide direct benefits to the female: (1) protection of females and their offspring from aggression by other baboons and (2) the development of close relationships with infants. Male protection and infant care may provide important rewards that motivate females to form friendships. Whether these benefits can also provide an *evolutionary* explanation for friendships, from the female point of view, is much harder to determine. At the very least, this chapter indicates some important consequences of friendship that could, in theory, have served as the basis for the evolution of friendship-forming tendencies among female baboons.

MALE PROTECTION OF THE FEMALE AND HER OFFSPRING FROM AGGRESSION BY OTHER BABOONS

In macaques and baboons, adult females and immature animals of both sexes often receive aid from others when they are being threatened or attacked (baboons: Kummer, 1968; Cheney, 1977a; Ransom, 1981; macaques: de Waal, 1977; Deag, 1977; Fedigan, 1976). In both genera, close relatives of both sexes are the most common providers of aid (baboons: Walters, 1980; Cheney, 1977a; Johnson, 1984; macaques: Kurland, 1977; Massey, 1977; Kaplan, 1977, 1978; Berman, 1980; Watanabe, 1979), but unrelated males have also been seen aiding females and immatures (e.g., Kaplan, 1977; Ransom, 1981). Seyfarth

(1978b) found that each of the two adult males in his troop of chacma baboons aided primarily those females with whom he had special bonds. This same pattern held in EC troop.

Male defense of a female and/or her juvenile offspring was scored whenever the following scenario was observed: (1) a female and/or juvenile was threatened, chased, or attacked by another troop member; and (2) a male chased, threatened, or attacked the aggressor, or the victim ran to an adult male and remained close to him for several seconds. In the latter instance, the male might or might not behave aggressively toward the opponent. Often, the male's proximity alone served to protect the victim. Defense that included aggression by the male toward the opponent was considered to be active defense; if the victim initiated proximity to the male and he did not threaten, chase, or attack the opponent, a passive defense was scored.

Male defense was observed 43 times during ad lib sampling, and 27 such events were observed during focal samples of females and males. Table 6.1 shows the breakdown of these events in terms of: (1) the age/sex class of the victim; (2) the age/sex class of the opponent; (3) whether the defense was active or passive; and (4) whether the defending male was a Friend or Non-Friend. In 93% of the ad lib events and 88% of the events observed during focal sampling, the defender was a Friend. In only one of the ad lib events, and in none of the focal events, was the opponent a Friend.

Since only 12% of all the possible adult male–female pairs in the troop were Friends, it is clear that Friends were disproportionately responsible for defense of females and their juvenile offspring. It is difficult, however, to attach a significance level to these results since it is not clear how expected frequencies should be determined. Females, and often their offspring as well, spent more time near Friends than near Non-Friends, and these males therefore probably were confronted with more opportunities to defend victims than were Non-Friends. It should be noted, however, that males could and did rush to the aid of victims who were dozens, or even hundreds, of meters away. Such long-range defense was possible because victims often stood bipedally or rushed about screaming, drawing attention to their plight, and these "displays" were usually easily detected by anyone in the troop. Detailed observations that included the number of opportunities to provide aid, estimates of how far away such opportunities could be detected, the proximity of different males to the victim, and the frequency with which aid was provided by different males might allow one to estimate how important proximity is in determining whether or not aid occurs and how quickly the aid is given. If proximity is important, then

Table 6.1. Male Defense of Females and Their Offspring: Friends and Non-Friends

	Ad lib observations (N=43)	Focal observations (N=27)
Defender		
Friend	40 (93%)	24 (88.9%)
Non-Friend	3	3
Opponent		
Adult Non-Friend male + Friend	1	0
Adult Non-Friend male	12	11
Adult Friend	0	0
Adult female	19	5
Juvenile	11	11
Victim		
Infant	5	5
Juvenile	6	4
Adult female	32	18
Type of defense		
Active	38	22
Passive	5	5
Effect of defense[a]		
Opponent ceases aggression		14
Opponent continues aggression[b]		10
Unable to determine[c]		3

[a]The effect of defense was scored only for events observed during focal samples since I was not always able to monitor ad lib events in sufficient detail to be sure of outcomes.

[b]In most cases the intensity of the opponent's aggression declined, probably in response to the defender's actions; in none of these cases did an opponent continue an *attack* following defense.

[c]In these cases, the female, defender, and opponent all disappeared into thick vegetation, and I was unable to determine the effect of the defense.

the greater likelihood of receiving prompt aid from a nearby male might be one of the reasons why females spent time near particular males.

Antigone and her Friend Sherlock are feeding about 10 m apart. Suddenly Handel, another adult male, runs toward them from over 20 m away and chases Antigone. She runs away, screaming and looking back toward Sherlock. Sherlock charges Handel, pant-grunting (an aggressive vocalization). Handel runs away from Sherlock and leaves the area. Sherlock sits near Antigone. She approaches him and begins to groom him. She grooms him for 10 minutes and then both resume feeding (focal sample on Antigone, 10 November 1977).

This sequence of events is typical of male defense of a female Friend against another baboon: The female was threatened or chased, the Friend threatened or chased the opponent, and the opponent stopped showing aggression. Sometimes the female solicited aid by running directly to her Friend, and in other cases the Friend rushed to help her without first being solicited. Grooming was not an inevitable part of the sequence, but females often groomed Friends after receiving help, and they sometimes used grooming as a way to solicit help. Grooming by the female of the defending male was observed in one-third of the observations in Table 6.1 that were derived from focal samples.

Importance of Male Protection

It is clear from Table 6.1 that male Friends provided nearly all of the aid females and young received from males. How important is such aid to a female's reproductive success? The answer depends, in part, on the identity of the aggressor and the context of the aggression. In EC, adult females and juveniles rarely harmed their victims during aggressive episodes, and they were never seen to inflict serious injuries. Male defense in such instances may have a small effect, but possibly a significant one when accumulated over time, by reducing the capacity of higher-ranking individuals to disrupt the maintenance activities of the victims (Altmann, 1980; Stein, 1981). Furthermore, support by adult males may sometimes be important in the rare instances where radical changes in female dominance relationships occur. In three different groups of macaques, for example, support of females by affiliated adult males was considered a critical factor in the females' successful attempts to rise in rank over previously dominant females (Chance *et al.*, 1977; Koyama, 1970; Gouzoules, 1980). During a similar shift in dominance relationships among EC females, however, adult males apparently played no role (Smuts, 1980).

When the opponent is another male, the value of a Friend's aid seems more obvious: Unlike females and juveniles, male opponents seriously injured EC females, and other males were the only allies large enough to provide a consistent deterrent to male aggression.[1]

[1] Occasionally, a whole group of females and juveniles will mob a male who has shown aggression toward a female or juvenile, and in these cases the

Thus, defense by Friends against other males may have an important effect on female reproductive success.

To test this hypothesis, we need to consider the costs to females of male aggression and the benefits to females of defense by Friends. Quantitative assessments of costs and benefits are not available, but they can be discussed in qualitative terms. The relevant data are reviewed below, but first I will briefly consider interactions between the two males when one male defends a female against another.

Behaviors of Male Opponents and Defenders

Although males may provide the female with the best protection from aggression by other males, from the male's point of view this type of defense is very different from defense against female or juvenile opponents. Any adult or subadult male can protect a female against another female or juvenile at small risk to himself, since subadult and adult males are larger and stronger than females and juveniles. When defending a female against another male, however, the defender faces someone who could injure him severely. This fact raises two questions. First, how risky is it for a male to defend a female against a male opponent? Second, can any male provide a female protection against other males or only those males who are dominant to the opponent?

It is not possible to measure risks directly, but a closer look at the responses of opponents to defenders may provide some clues. In nearly one-half (48%) of the 23 observations of male defense of a female or her offspring against a male opponent, the defender did not interact directly with the opponent, and his proximity alone usually resulted in cessation of aggression. In another 39% of the observations, the defender threatened or chased the opponent, and the opponent moved away without further aggression. In only 3 of the 23 interactions did the opponent and defender fight; none of these fights was prolonged, and none resulted in injury to either male. These 3 fights accounted for less than 5% of all fights observed among males.

These results suggest that defense of a female against a male opponent is often not very risky, because the opponent usually retreats without escalating. This was true even when the defender was subor-

---

smaller animals are able to provide an effective deterrent to male aggression. Mobbing, however, is not a common response to male aggression. It was used mainly against males who had recently entered the troop, and it was more common in response to aggression against an infant than in response to aggression against an older individual (mobbing is discussed further in Chapter 8).

dinate to the opponent. For 17 of the 22 observations, it was possible to judge the dominance relationship between the opponent and defender based on the outcome of other agonistic interactions observed between them (see Chapter 7). In 11 of 17 cases (65%), the opponent was a male who won at least one other encounter with the defender, yet in all but two of these cases the opponent retreated without escalation.[2] These results indicate that even a male of lesser competitive ability than the opponent can provide a female with protection. Presumably, a smaller or less dominant male is able to provide effective aid because he can increase significantly the costs to the aggressor of continuing to threaten or attack the female (in terms of energy expenditure and risk of injury), even though he is not likely to win a fight with the opponent. If the benefits to the opponent of continuing to show aggression toward the female are smaller than the costs of confronting the defender, the opponent should retreat, even if he is likely to win a fight with the defender, since in any fight between subadult or adult males, both parties risk injury. This scenario is best illustrated by numerous observations from EC of large juvenile males who, individually, effectively defended females or juveniles against fully adult males. In these cases, the defending male was only slightly larger than an adult female and probably weighed only a little more than one-half of what his opponent weighed; his canines were larger and sharper than a female's but no match for those of an adult male. These young defenders were always putative relatives of the female (i.e., her son) or the juvenile (i.e., a maternal sibling) being protected. The same sort of fierce defense was typical of the young adult males who were putative sons of their female Friends. These males were more willing to come to the aid of their Friends (and their Friends' offspring) than were other Friends: Putative sons accounted for only 6.5% of all Friend dyads, but they were responsible for 19% of the observations of defense of females and offspring by Friends.

This observation raises an important point: Males did not always aid their Friends when opportunities arose. Sometimes males ignored a Friend's plight and continued their current activity. On a few occasions, I saw a male walk slowly and deliberately away as soon as his Friend was threatened or chased by another male, as if to avoid being confronted with the opportunity to come to her aid. Unfortunately, I recorded systematically only those cases in which the male did aid a female and not the total number of opportunities to provide defense. A

[2] In one case, the opponent attacked the defender; in the other case, the defender attacked the opponent.

study of the frequency with which different males responded to such opportunities would help to provide a more accurate assessment of the risks a male incurs when defending a female against another male, the extent to which males vary in their willingness to expose themselves to these risks, and the factors affecting this variation (e.g., kinship, length of friendship, age, dominance).

FRIENDSHIP AND MALE AGGRESSION TOWARD FEMALES

Frequency and Severity of Aggression

During focal samples, females were victims of male aggression on average once every 17 hours.[3] This translates into about five aggressive episodes per week, per female, if we assume that females did not receive aggression from males after dark. (This assumption is no doubt invalid, but I have no way of estimating rates of aggression at night. The estimates that follow are, therefore, conservative; they indicate *minimum* rates of male aggression toward females.) Of these aggressive episodes, 25% included attacks; the rest involved threats and chases only. Thus, on average, each EC female was attacked by a male slightly more than once a week—at the very least.

How severe were these attacks? The answer is based on the frequency of serious wounds among adult and adolescent females derived from daily censuses.[4] A serious wound was defined as a cut at least 3 cm long and deep enough to cause a gaping, bloody gash [the most common wounds, cuts on the perineum, were excluded since they tended to be superficial and to heal quickly (Hausfater, 1975)]. Most of

[3]A total of 241 hours of focal samples on males, collected over a 3-month period, provided a similar estimate of rates of aggression. Males showed aggression toward females once every 7 hours. When this value is multiplied by the number of males (18) and divided by the number of females (36), the expected frequency of aggression per female is once every 14 hours.

[4] I assume that all serious wounds resulted from attacks by adult males for two reasons. First, I observed adult males inflict serious wounds on females twice, but I never saw such wounds inflicted by another female even though I observed about five times as many attacks on females by other females as I did attacks on females by males. Second, all of the serious wounds observed among females conformed to the characteristic appearance of cuts inflicted by the canines of an adult male. I was familiar with the appearance of such wounds because they were a common result of male–male aggression. The only other possible source of wounds is a large predator, such as a leopard or lion. These animals were extremely rare, and if a female were seriously injured by a large cat, it seems unlikely that she would escape.

these gashes appeared on the legs, arms, or shoulders. A few were on the side of the body, and only one was on the head. (In contrast to males, females rarely received wounds on the face, probably because, unlike males, they did not turn to face their opponent but instead crouched immobile on the ground during an attack.) Many of these wounds were 10–15 cm long and clearly quite painful, causing the female to limp for a week or more. Several wounds took 2 or 3 months to heal completely, and one female, an adolescent, died from a long gash on her abdomen that exposed her entrails. Again, estimates of the frequency of wounds are conservative, since some serious wounds might have gone undetected because they were hidden by thick fur.

A serious wound was observed among EC females once every 6.8 days of observation—just about once a week. Since each female was attacked about once a week, this means that, of all the females attacked in a given week, on average one received a serious injury. Based on male focal samples, the wound data imply that 1 out of every 53 attacks resulted in serious injury to the female (this explains why we actually saw males inflicting wounds on females only twice during the 1977–1979 study period). Each EC female, then, could expect to receive at least one serious wound from a male every year.

Contexts of Male Aggression Toward Females

In Table 6.2, 93 cases of male aggression toward females observed during focal samples of females and males are classified in terms of the male's and female's behaviors just prior to the aggression. These contexts were defined as follows:

Apparently Unprovoked Aggression. Neither the male nor the female were interacting with each other or with another baboon just before the aggression, and there was no apparent reason for the male's behavior.

> *Jocasta is feeding on grass. Handel, an adult male, is also feeding, 25 m away. He looks at Jocasta, runs toward her, jumps on her, and bites her back. She screams and runs away. He chases her briefly and then returns to his previous feeding site (focal sample on Jocasta, 15 May 1978).*

This was the most common "context" of aggression by males toward Non-Friend females (see Table 6.2).

Feeding Competition. The male and female were feeding close together, and he appeared to use aggression to force the female to "keep her distance" or to abandon her feeding site to him.

Table 6.2. Percentage of Male Aggression Toward Females in Different Contexts[a]

Context[b]	Male aggression toward females (%)			Attacks (%) (expected value = 25%)[c]	Number of aggressive episodes
	All combined	Friends (N=20)	Non-Friends (N=73)		
Apparently unprovoked	23.7	15.0	26.0	40.1	22
Feeding competition	20.4	25.0	19.2	15.8	19
Redirected	18.3	5.0	21.9	11.8	17
After unprovoked female submission	8.6	5.0	9.6	12.5	8
Immediately after another male shows aggression toward same female	7.5	10.0	6.9	28.6	7
In defense of a third party	7.5	5.0	8.2	28.6	7
Female begins to leave close proximity	3.2	15.0	0	66.7	3
Miscellaneous	2.2	5.0	1.4	4.4	2
Unable to determine context	8.6	15.0	6.8	13.4	8
Total	100	100	100		93

[a]All observations derived from focal samples on females (N=926 hours) and males (N=241 hours).
[b]See text for explanation of contexts.
[c]Over all categories, 25% of the aggressive episodes involved attacks.

Helen feeds on grass. Pliny, a subadult male, approaches to within 2 m and raises his brows at her (a threat). She fear grins and moves away. He takes over her feeding site (focal sample on Helen, 5 June 1978).

These interactions were very similar to those observed between females in similar contexts. Feeding competition was the most common context of aggression by males toward female Friends.

Redirected Aggression. The male was involved in aggression or a tense interaction with another male when he chased, threatened, or attacked the female. Males often seemed to use redirected aggression not only to "vent frustration" but also as a means of extricating themselves from a difficult situation without "losing face."

Virgil is digging for corms. Triton approaches, stands right in front of him, and waits, staring at Virgil's feeding site. Virgil stops feeding, gazing around in all directions (except at Triton). Then Sophia walks by. Virgil charges her, and they disappear into the bushes. Triton feeds at Virgil's spot (focal sample on Virgil, 18 October 1978).

Redirected aggression is common among other age/sex classes of baboons and in many other nonhuman primate species (e.g., Goodall, 1975; de Waal, 1977).

Unprovoked Female Submission. As noted in the previous chapter, a female sometimes showed fearful or submissive responses when in close proximity to a male, even if the male had not directed any aggression toward her. Sometimes these indications of nervousness appeared to provoke an aggressive response.

Zandra is feeding. Handel, an adult male, walks slowly toward her, foraging as he moves. Zandra darts several quick glances in his direction, and, as he approaches, she jumps away from him with a loud geck. Handel attacks Zandra and she runs away, screaming. Handel resumes foraging (focal sample on Zandra, 7 November 1977).

Immediately After Another Male Shows Aggression Toward the Same Female. On a few occasions, one male threatened, chased, or attacked a female, and a few seconds later another male also showed aggression toward the same female. In all cases, the first male stopped showing aggression after the second male entered the interaction, and I had the impression that the second male actually might have intervened on the female's behalf. Male use of aggressive behaviors to defend females is considered further when I discuss "mock" attacks, below.

Defense. This context for male aggression has already been discussed: Males sometimes threatened, chased, or attacked females who were showing aggression toward one of the male's female Friends or her offspring.

Female Leaves Close Proximity to the Male. The male and female have been feeding in close proximity for several minutes. The female then begins to leave, and the male shows aggression toward her. All three observations of aggression in this context involved Friends. These incidents were difficult to interpret. Among hamadryas baboons, which live in one-male, multi-female groups, the male routinely shows aggression toward "his" females when they move away from him, and in this species females typically respond to such aggression by moving closer to the male (Kummer, 1968). However, in savannah baboons, females tend to move away from a male aggressor, and males instead tend to use affiliative behaviors to encourage females to remain in proximity. It is possible that in these three cases the male was showing hamadryas-like behavior, but the female in each case did run away. More likely, the male showed aggression toward the female not because he did not want her to leave, but because her sudden movement startled him.

Miscellaneous. Two contexts for male aggression were observed only once each. In one case, a male (Non-Friend) chased a female who was presenting to another male. In the other case, a male showed aggression toward a female Friend who was soliciting his aid against another male who had just chased her. The Friend turned away when the female ran over to him, and when she then moved around to face him, he attacked her.

Not Known. For various reasons, I was sometimes unable to determine the context of male aggression toward females, for example, when the episode began while the animals were briefly out of view in the bushes. This context differs from the first category, unprovoked aggression. In the latter case, I knew the behavior of both male and female before the aggressive incident but could detect no basis for the male's behavior.

Aggression by Friends

The results in Table 6.2 suggest the possibility that Non-Friend males and male Friends tended to show aggression toward females in different contexts (e.g., Non-Friend males showed unprovoked aggression relatively more often, and Friends showed aggression during feeding competition relatively more often), but the sample sizes for Friends are too small to permit a quantitative comparison.

Over all contexts, Friends were responsible for 21.5% of male aggression toward females. This was significantly *more* aggression than expected based on the proportion of Friend dyads in the troop (12%) [*1*] but significantly *less* aggression than expected if we assume that the frequency of aggression was a direct function of the amount of time males spent within 15 m of females [*2*]. Time spent in proximity probably did affect rates of aggression, but the number of aggressive episodes observed during focal samples is too small to provide an estimate of how strong this relationship was. It is not possible, therefore, to say with any certainty whether Friends contributed more or less aggression than expected, but the fact that they were responsible for about one-fifth of all aggressive episodes indicates that females were by no means immune to threats, chases, and attacks from their male Friends.

Perhaps of greater interest, from the point of view of long-term effects on female survival and reproductive success, is the intensity of aggression. Surprisingly, Friends were perpetrators of the most severe form of aggression, attacks, more than expected based on their participation in all types of aggression (see Table 6.3), but this result was not significant [*3*]. Once again, however, the results suggest that being Friends with a male did not necessarily protect a female from aggression by him.

Why do males *attack* females? Since the difference between the expected and observed number of attacks for different contexts of aggression is small in most cases (Table 6.2), the context of the aggression does not appear to predict the intensity of aggression. There

Table 6.3. Percentage of Chases, Threats, and Attacks of Females by Male Friends

Intensity of aggression	Percentage by Friends	Number of observations (Friends and Non-Friends combined)
Threats	15.4	26
Chases	20.5	44
Attacks	30.4	23
All	21.5	93

is one possible exception: The observed number of attacks was considerably (though not quite significantly) higher than expected in the context "unprovoked aggression" [*4*], but this finding seems only to deepen the mystery.

Although the quantitative data fail to resolve this question, qualitative observations can provide further insight into male motivations for attacking females. These observations are discussed below at some length because a better understanding of male aggression toward females is crucial to an assessment of the costs and benefits to females of friendship with males.

Qualitative Aspects of Male–Female Aggression

Prolonged, "Unprovoked" Attacks. I noted above that attacks on females were more common in the context "unprovoked aggression" than in other contexts, and nearly all of the truly severe attacks that I observed occurred in this context. These involved a prolonged, seemingly obsessive series of attacks on a female who, by the end of the incident, was often literally shaking with fear and choking on her own screams. In one case, I saw the female crawl under a bush, shuddering, after one of these persistent attacks. I went to look for her 20 minutes later because the troop had moved on. I found her still huddled under the bush, her head buried in her arms. As far as I could tell she was not injured in the attack—just very frightened.

These prolonged attacks puzzled me for a long time. Why should an apparently unprovoked male expend so much effort in order to terrify a female one-half his size? I sometimes wondered if perhaps these incidents were simply irrational acts by males who had temporarily "lost control of their emotions"—a disquieting and unsatisfying conclusion for someone who expects the regular behavior patterns of wild animals to make sense. Then I observed an incident that provided an important clue to this puzzle.

> *Phaedra is feeding on grass, surrounded by other baboons doing the same. Adonis, in consort with Andromeda, feeds 20 m away. He looks up at Phaedra and stares at her for a few seconds; she is unaware of his attention. Suddenly he rushes over and, with his canines, tears an 8 cm gash in her arm. Phaedra screams in pain and collapses on the ground. Adonis saunters back to Andromeda. (Phaedra's wound took 2 months to heal fully, and she did not use her arm for over 1 week) (ad lib observation, 16 November 1978).*

This is a good example of an "unprovoked attack," although in this case the attack was not prolonged, perhaps because the male managed to inflict a serious wound at the start. However, I had been observing Adonis, Andromeda, and Phaedra off and on for several days, and these observations provided a possible reason for the attack. Adonis and Andromeda were Friends who had been in consort for over 1 week. He was being an unusually attentive partner, grooming her frequently and following her closely. Phaedra, a primiparous female who was also a Friend of Adonis, seemed annoyed by his attentions to Andromeda, who ranked below her, and she repeatedly harassed Andromeda while the latter was in consort with Adonis. I saw Phaedra chase Andromeda away from Adonis several times in the 3 days preceding the attack. Each time, Adonis did what males in consort are forced to do if they want to avoid losing the female to another male—he ran after Andromeda. This meant that he had no opportunity, on these occasions, to show aggression toward Phaedra. When he did attack her, it seemed plausible that he was punishing Phaedra for her harassment of his consort partner. It worked—Phaedra, with her severe wound, ceased to show aggression toward Andromeda.

My conclusion that Adonis attacked Phaedra because of her aggression toward Andromeda was based on observations of incidents that occurred more than 24 hours apart; it seemed like a logical inference, but was by no means the only possible explanation. A second incident observed more recently has convinced me that baboons are indeed capable of delayed "revenge" or "punishment."

Pegasus, a large subadult male, sits resting in a patch of early morning sun on top of a termite mound. He is at the center of a cleared circular area about 30 m in diameter. His closest Friend, Cicily, sits next to him, holding her black infant. (These two, who were about the same age, grew up together in EC; they were probably not closely related. They had one of the strongest bonds I have ever observed between a male and female baboon. Cicily, a primiparous female, was unusually small for a baboon mother, and this might have been one reason why Pegasus was so solicitous of her welfare. See Figure 6.1.) Cicily wanders off and begins to forage at the edge of the cleared area. Zora, a high-ranking adolescent female, chases Cicily. Pegasus, observing this incident, grunts and looks as if he is about to intervene when Zora and Cicily disappear into the surrounding bush. Pegasus resumes his tranquil pose. We, the observers, remain in

Figure 6.1. Pegasus grooms his best Friend, Cicily, who is nursing her first infant.

the cleared area. After 10 minutes, Zora appears at the edge of the clearing—this is the first time she has been visible since she chased Cicily. Pegasus reacts instantly. He leaps off the mound and pounces on Zora, emitting a series of aggressive "hum-roar" grunts. He repeatedly grabs her neck in his mouth and lifts her off the ground, shakes her whole body, and then drops her to the ground. Zora screams continuously and intensely and keeps trying to escape. Each time Pegasus catches her and continues his attack. This goes on for a full 5 minutes. I am struck by the vehemence of the attack and so, apparently, are the females and juveniles nearby (no adult males are in view); they have all ceased foraging and are staring at Zora and Pegasus. Pegasus finally releases Zora. She has a minor cut on her perineum and a more severe gash on the palm of her hand. She limps for several days (ad lib observations by John Watanabe and the author, 12 June 1983).

There seemed to be no reasonable explanation for Pegasus's behavior except in terms of the chase he observed between Zora and his Friend, Cicily. His actions once Zora reappeared were typical of the prolonged "unprovoked" attacks I described above. I conclude that some—perhaps most—of these "unprovoked" attacks are just the opposite: They are attacks in response to specific acts by the victim that the attacker observed and remembered; it is the observer who does not know the reason for the male's behavior, not the baboons. It is indeed possible that baboons remember such incidents and act on the basis of these memories not only over periods of 10 minutes, as Pegasus did, or even hours, as Adonis did, but for days, weeks, or even longer. If I am correct that many "unprovoked" attacks do represent delayed punishment for aggression shown toward a Friend of the male attacker, then defense of Friends is a far more important context for male aggression, and particularly for intense aggression, than the data in Table 6.2 would indicate.

The Victims of Redirected Aggression. In the three incidents described below, baboon A was for some reason unwilling to show aggression toward baboon B, the "target," either because B was an adult male or because B was being defended by an adult male. Baboon A instead showed aggression toward C, who was a Friend or relative of B. Other baboons then became involved—either as defenders of C or as additional victims of redirected aggression. In all three cases, A chose C, a Friend or relative of B, out of a large pool of possible victims.

> *Antigone, an adult female, attacks Artemis, another adult female. Artemis runs away, screaming with her tail raised. Hector, a Friend of Artemis, chases Antigone. Sherlock, a Friend of Antigone, threatens Hector but Hector ignores him. Sherlock chases Circe, the only other Friend of Hector's nearby. Hector continues to chase Antigone. Antigone runs to Sherlock, who has stopped chasing Circe, and Hector chases Antigone's immature son (ad lib observation, 17 August 1978).*

> *Boz attacks a juvenile female, Pyrrha. Pyrrha's subadult brother, Plutarch, charges over from 50 m away and immediately attacks Leda, the only female nearby who is a Friend of Boz. Leda runs to Ovid, a subadult male Friend. Plutarch then chases Minerva, Ovid's only other female Friend in the vicinity (focal sample on Joseph, 25 June 1983).*

> *Archimedes, a subadult male, chases Andromeda, an adult female, who runs over to Sherlock and grooms him. Archimedes*

looks around at the 30 or so baboons nearby and spots Caesar, Andromeda's juvenile brother. He chases Caesar. Caesar runs away, and Archimedes looks around again and spots Sparta, Andromeda's younger sister. He threatens her, but Sparta, sitting next to her Friend Bacchus, ignores him (ad lib observation, 7 July 1983).

One-third of the observations of redirected aggression in Table 6.2 fit this pattern: The male redirected aggression toward a female Friend of the male with whom he had been interacting just prior to the aggression. Similar patterns of redirected aggression toward relatives of the target have been reported in macaques (Judge, 1982).

There is a certain familiar ring to these tactics. People, of course, redirect aggression in this way all the time—if we cannot get B, then we go after someone who means something to B. From a sociobiological perspective, such tactics make sense. Because aggression toward a relative of B inflicts a cost on B, A can use such aggression to manipulate B's behavior; a Friend of B can be used in a similar way if B's relationship with that Friend is important to B's own fitness. The fact that baboons redirect aggression to Friends of the target shows that they are aware of the affiliative relationships of other troop members, and that they are able to use this knowledge to manipulate the outcome of an encounter.

These findings indicate that friendship with a male may involve costs as well as benefits for the female. Simply because of her close relationship with a particular male, she may become the victim of aggression by other males. In many cases, this aggression occurs in contexts like those described above and falls into the category, redirected aggression. In other cases, aggression toward a female appears to function as a direct challenge to the female's Friend.

Phoebe and her Friend, Cyclops, are resting together. Virgil, an adult male, approaches and attacks Phoebe. During the attack, they move away from Cyclops. Cyclops follows them, pauses about 5 m away, and stares at Virgil. Virgil ends his attack, moves close to Cyclops, and threatens him by yawning in his face with exposed canines. Cyclops remains in place, staring at Virgil, but not responding to his threats. Virgil makes a rapid movement toward Phoebe, Cyclops rushes after him, Virgil turns to face Cyclops, and they fight. Phoebe disappears into some nearby bushes. The fight ends with Virgil chasing Cyclops into the bushes. As soon as they disappear, Phoebe emerges from the shrubbery and resumes feeding (focal sample on Phoebe, 18 May 1978).

In this example, it seemed as if Virgil used the attack on Phoebe to initiate a fight with Cyclops—otherwise, why attack Phoebe when Cyclops was nearby? Cyclops was the most successful male in the troop in terms of consort activity and access to meat, and perhaps as a result he received many challenges to fight from other males. He consistently ignored these challenges, and it was therefore very difficult for other males to "test" his competitive ability (see Chapter 7). By attacking Phoebe, Virgil put Cyclops in a difficult position: If he ignored Virgil's behavior, he would also be allowing his closest female Friend to be attacked and perhaps injured right before his eyes. In this case, the challenge worked: Cyclops responded to Virgil, and Virgil "won" the fight (i.e., he was able to chase Cyclops away).

During several other instances of "unprovoked" aggression, I also had the impression that the female was being used as a means of initiating an interaction with her male Friend. Young adult natal males and recent immigrants seemed particularly likely to use this technique. Recently matured males and new immigrants are faced with the problem of evaluating the competitive abilities of other males in relation to their own. In theory, a male could do this by initiating fights directly with established males, but this tactic has at least two disadvantages. First, it is costly, since fights are potentially risky for both winners and losers. Second, residents tend to ignore young males' attempts to initiate agonistic encounters (Smuts and Watanabe, in preparation). By showing aggression toward female Friends of older males and evaluating the males' responses, young males can probably learn a lot about the personalities and competitive abilities of their rivals. Depending on the resident males' responses, this aggression might sometimes, paradoxically, improve the aggressor's chances of forming bonds with his female victims. If the female's current Friends repeatedly fail to defend her effectively against the newcomer's aggression, it may be in her interest to form a new friendship with the immigrant who has so vividly demonstrated his superior competitive ability. This last suggestion is pure speculation, but it is intriguing because it could help to explain why females sometimes seemed attracted to males who had been aggressive toward them (note that this was not always true; females sometimes developed strong aversions toward aggressive males). Similar reasoning has been used to explain the puzzling tendencies of female gorillas to leave their group to join another male after that male has succeeded in invading her old group and killing her infant: If the invader's infanticide means that he is a stronger male than the leader of the group she is in, then it might be in the female's long-term interests to join this new male (Wrangham, 1979a; Fossey, 1983).

Mock Attacks and Use of Females as Agonistic Buffers. During what I call "mock attacks," a male goes through the motions of attacking a female, but several characteristics of the incident give the impression that the attack is not genuine, in the sense that the male does not intend to frighten or injure the female. Most mock attacks involved male Friends. In some of these cases, it seemed as if the Friend conducted a mock attack on the female as a way of defending her against another male.

> *Aristotle, a subadult Non-Friend, chases Andromeda. She screams and runs to Adonis, her Friend. Adonis jumps on top of Andromeda, who crouches beneath him, and places his open mouth on her back, as if to bite her. Andromeda is silent. Aristotle watches Adonis and Andromeda for a few seconds and then wanders off. Adonis releases Andromeda and immediately presents for grooming. She grooms him (focal sample on Adonis, 8 December 1978).*

In this case, I was close enough to see that Adonis was not actually biting Andromeda's back—his lips covered his teeth and he did not press down hard on her back. Several other features of this incident led me to interpret his behavior as defense rather than aggression: Aristotle's prior aggression toward Andromeda, her solicitation of aid, her failure to scream or show other signs of fear while Adonis "attacked" her, and the fact that she calmly groomed him afterward rather than running away.

But why should a male conduct a mock attack on a female as a means of defending her? During a mock attack, the male is literally in possession of the female. This may function as a particularly strong message to the opponent: "This is my female; if you show aggression toward her you have to deal with me." When a mock attack is underway, the opponent cannot continue to show aggression toward the female without also showing aggression to the male Friend; the two are so closely bound up that to threaten or attack one implies a threat or attack to the other. When a mock attack was used in this way, the female seemed to understand her Friend's intent: She did not struggle or vocalize but submitted passively to the charade.

As described so far, mock attacks are just a peculiar form of defense and not really aggression at all. The situation is more complicated, however, because in other cases the benefit to the female of the mock attack was less clear, and it appeared that the male was using her to mediate an interaction with another male. In these cases, the male's and female's behaviors were also more ambiguous, and elements of mock attacks and real attacks were often combined.

Handel is eating meat. Cyclops and Dido sit together, just a few feet away. Dido is a Friend of both males. Cyclops repeatedly "attacks" Dido. Each time Dido grimaces and squeaks mildly, but as soon as Cyclops releases her she continues to sit near him. During the attacks, Handel's attention is drawn away from the meat and toward Dido and Cyclops. Handel appears increasingly upset by these attacks. Finally, during one of them, he chases the juvenile offspring of one of Cyclop's other Friends, abandoning the prey. Cyclops immediately takes the meat (focal sample on Cyclops, 12 November 1978).

Strum (1983) has identified a different but possibly related pattern of male "use" of females in the adjacent Pumphouse troop. In such incidents, which I also observed in EC, a male sought close contact with a female during a tense or aggressive encounter with another male. The female, who was usually a close affiliate of the male, was approached with exaggerated lip-smacking and friendly grunts. The male then attempted to groom her or simply to clasp her body to his. Strum reports that this use of females as "buffers" was often effective (i.e., the interaction between the two males was altered in a way that appeared to benefit the male using the female). One reason female buffers are effective, Strum argues, is that the opponent must now show aggression toward the female if he continues to show aggression toward the other male, and aggression toward a female may result in mobbing by other troop members; this is particularly likely if the aggressor is a recent immigrant. I made a similar point above, but with the male and female roles reversed, when I argued that an opponent's aggression toward a *female* might be inhibited by physical contact between another male and that female during a mock attack. There is no reason why this tactic cannot work both ways, depending on the context and the identities of all three participants. The essential point is that by joining their bodies together, a male and a female can increase the costs to the opponent of further aggression above what they might be if he were free to pursue either party separately.

Strum (1983) has emphasized the importance of female cooperation during these buffering incidents, and female cooperation also seemed to be important during mock attacks. If the female refused to cooperate and instead vocalized, struggled, and tried to flee, the male sometimes showed genuine aggression toward her, presumably in an attempt to punish her or to force her to cooperate. I have seen males attack females who refused to cooperate during both buffering and mock attack incidents. The distinctions, then, between a male's use of a female Friend as a buffer, mock attacks by Friends, and "genuine" aggression by Friends are sometimes not clear, and I would argue that it is best to view these behaviors as points on a continuum rather than as discrete types of interactions.

It is also possible that some of the aggression observed between Non-Friend males and females is based on similar motivations, for example, attempts by males to use, or even defend, females, either by initiating close contact or through mock aggression, which turn into genuine aggression because of the female's response to the male's behavior. As noted in the previous chapter, female reactions to male approaches are highly variable, and, not surprisingly, their responses to less neutral behaviors by males (e.g., buffering attempts or mild aggression) also vary tremendously. Much of this variability relates to the length of the male's residence in the group and whether or not he is a Friend of the female. For example, I saw a female respond to a mild threat from a male over 30 m away with prolonged screaming and trembling; the male in this case was a recent immigrant. On another occasion, I observed a female completely ignore a male careening down a hill straight toward her, pant-grunting and threatening; in this case, the male was a Friend. These examples, along with the female responses to close approaches involving Non-Friend males described in the previous chapter, indicate the difficulty that Non-Friend males, especially recent immigrants, had in achieving relaxed proximity to females. I suspect that some male aggression toward females reflects the male's frustration at the female's refusal to let him come close and to let him "use" her in ways that she allows her Friends to do.

Summary and Discussion

The quantitative and qualitative results discussed in this chapter suggest that although Friendship with males provides females with an important benefit—defense against aggression by other baboons—these relationships may also impose costs. The potential disadvantages of friendship include becoming a focus of redirected aggression by a male involved in an agonistic encounter with the Friend, being used by another male as a means of directly challenging the Friend, and, finally, being used by the Friend (either through mock attacks or as a buffer) to manipulate the outcome of a tense or aggressive interaction with another male. I have emphasized these costs not because I believe that they are so great as to outweigh the benefits of defense by Friends but to stress the multiple, complex, and interrelated sets of social tactics and social relationships that must be considered when trying to answer what might at first appear to be a relatively simple question: Why do females form friendships with males?

The preceding analysis suggests that females form friendships in part because of their extreme vulnerability to males in general. A female baboon is continuously surrounded by males who could quite easily injure, or even kill, her or her offspring. Her most effective defense against these dangers is another male. But rather than relying on the possibility that *some* male will risk injury and come to her aid, the female forms a long-term bond with one or two males who are more reliable allies, presumably because they in turn receive benefits from their relationship with her. These benefits will be discussed in Chapter 8; the important point here is the possibility that friendship has evolved as a result of female vulnerability to aggression from other troop members and perhaps, especially, from males.

AGGRESSION TOWARD INFANTS

Because of its helplessness, the female's infant is probably even more vulnerable to aggression from other baboons than is the female herself. Severe harassment of infants and young juveniles by adult females has been reported in one troop of wild yellow baboons (Wasser, 1983; Wasser and Barash, 1983) and among captive bonnet macaques (Silk, 1981c). In a few cases, such harassment might have caused the death of the immature victim. Although EC adult females frequently showed aggression toward young baboons, severe harassment and injury to immatures was not observed. However, in some troops, male protection of offspring against harassment from adult females might be the most important benefit male Friends provide females.

Male protection of offspring against aggression from other males is probably another very important benefit of friendship to females. There is increasing evidence that, in a wide variety of animals, including many non-human primates, males that have recently entered a group sometimes kill infants fathered by other males (see Hrdy, 1979; Hrdy and Hausfater, 1984; and Leland *et al.*, 1984 for reviews). After the death of an unweaned infant, females in many species resume sexual cycles and conceive earlier than they do when infants survive, and so infanticide, it is argued, may increase mating opportunities for the infanticidal male (Sugiyama, 1965; Trivers, 1972; Hrdy, 1974).

The hypothesis that infanticide represents an evolved male reproductive strategy is supported by direct evidence of infanticide by males unlikely to have fathered the infants they kill and by evidence that these males often subsequently mate with the mother (see Hrdy [1979] and Leland *et al.* [1984] for reviews). It is also supported by a wide variety of behaviors by mothers, probable fathers, and infants that appear to represent "counterstrategies" for reducing male infanticide (see works cited above). In grey langurs, for example, mothers have been observed to leave the group temporarily after immigration by a new male (Hrdy, 1977). The evidence for male infanticide and female counterstrategies in savannah baboons is reviewed below.

Male Infanticide in Baboons

The best evidence for male infanticide comes from two long-term study sites: Gombe (olive baboons) and Moremi (chacma baboons) (Collins *et al.*, 1984). Data were available for four troops of Gombe baboons over a 10-year period. During this time, researchers observed four fatal attacks on infants by adult males. Twelve other infant deaths were associated with wounding, and the authors concluded that in at least five of these cases the wounds had almost certainly been inflicted by other baboons, probably by adult males.

In two troops of chacma baboons in Botswana studied by Busse, two infants were killed by adult males, and two others died after receiving wounds of unknown origin (it is important to mention that one of the infanticides occurred under artificial circumstances after the observers had tranquilized a mother with a blow-dart; females with infants were no longer darted after this event) (Collins *et al.*, 1984). Two other infants survived attacks by adult males.

In seven of the eight attacks at Gombe and Moremi in which the identity of the attacker was known, he was a male not present in the group when the infant was conceived. In the eighth case, the infanticidal male was still living in his natal group, but his mother and most of his siblings had left the troop when it divided 16 months before. Although he was in the troop when the infant was conceived, he had not been observed consorting with the mother.

There is also evidence that male infanticide occurs among Gilgil baboons (Nicolson, 1982; Smuts, 1982). Zephyr, the 4-month-old daughter of a high-ranking female, Zandra, became separated from her mother in dense vegetation. While Zandra was searching frantically for her daughter, we heard sounds of males fighting and an infant screaming. Zandra rushed toward the sounds. By the time we arrived a few seconds later, Zandra was cradling her unconscious infant who had deep puncture wounds in the skull and inner thigh. No other baboons were near Zandra and the infant when we arrived. The infant did not regain consciousness and died a few hours later. There was no evidence to indicate which male bit the infant, but since no non-troop members were seen near EC at any time that day, it seems likely that the attacker was a troop member. (An attack by a predator is very unlikely; leopards are extremely rare and very shy of humans in this area, and dogs, which occasionally attack baboons, were never seen without their human companions, who always made themselves known to us. In addition, the shape and depth of the wounds were more consistent with adult male baboon canines than with the much smaller canines of a dog.)

Several other cases of observed or suspected infanticide are summarized in the report discussed above (Collins *et al.*, 1984), and another case, committed during an intergroup encounter, is described by Shopland (1982).

Observations from Amboseli suggest that recently transferred males may endanger infants even before they are born (Pereira, 1983). A strange male, Kong, transferred into the study group. Within a few days, the top-ranking male and 5 of the 10 fully adult females received severe injuries. Based on long-term observations of this group and on Kong's frequent aggression toward troop members, the observer concluded that Kong had probably inflicted these wounds. Shortly afterward, three of the five pregnant females aborted, and one of the four infants in the troop died. The cause of death was not known, but the infant's mother was one of the females who had been attacked. Miscarriages are very rare among Amboseli baboons (only 3 other cases have been detected in 83 pregnancies), and so it is likely that the fetal deaths were a result of Kong's disruptive presence and/or attacks. As a result of their reproductive losses, all four females conceived considerably sooner than they would have had their infants survived. It is not known whether Kong consorted with any of these females, but he was the highest-ranking male in the group when they conceived, and, at Amboseli, high-ranking males often have high consort success (Hausfater, 1975).

These observations suggest that immigrant males may attempt to increase their mating opportunities both through direct attacks on young infants and through harassment of pregnant females who subsequently abort. The second phenomenon has been reported in wild horses, in which a number of fetal deaths occurred in association with male immigration and aggression toward pregnant females (Berger, 1983). It is possible that in baboons, male attacks on pregnant females and subsequent fetal losses are aberrant behaviors. However, these observations are also consistent with the hypothesis that natural selection has favored a tendency in male baboons to use aggression toward females and infants as a means of increasing their own reproductive opportunities, *whenever circumstances permit.*

Infanticide and induced abortions are unusual events, and they are subject to all of the problems of interpretation associated with any rare behavior. There is another source of information relevant to this hypothesis, however: the routine behavior of females, infants, and other troop members who have a stake in the welfare of those females and infants. Baboons exhibit a wide variety of behaviors that indicate that they view males—and particularly recent immigrants—as dangerous interlopers whose very presence threatens mothers and infants. These behaviors are described below.

1. Infants and young juveniles sometimes show intense negative responses to unfamiliar males. These reactions include screaming when the male is nearby and also approaches involving "simultaneous presenting, threatening, defecating and vocalizing; this behavior is referred to as 'ambivalence' " (Packer, 1979a, p. 27). Immatures show screaming and ambivalence significantly more often to recently transferred males than to more familiar ones (Packer, 1979a; Busse, 1981). Resident males frequently charge toward newcomers that have elicited fear in youngsters, and the strange males usually run away from these charges.

2. Mothers of unweaned infants frequently show alarm by raising their tails and/or screaming in the presence of recent immigrants. Busse (1984b) recorded mothers' responses to males whenever the distance between them was less than 5 m. Tail-raising and screaming were directed almost exclusively toward males that had transferred into the group *after* the mother conceived; such responses were not shown to males who could have fathered the infant, and they were never shown by females not carrying infants.

3. Females sometimes indicate their fear of males by avoiding them before the males have an opportunity to get close. During female focal samples, I recorded the female's response each time a male moved to within 5–15 m of her. If the female immediately moved away from the male, this was scored as a long-distance avoid. Movements by recent immigrants elicited a long-distance avoid by pregnant and lactating females more than twice as often as movements by resident males (median rate of avoids for short-term residents and newcomers combined: 15%; for long-term residents: 6%) [*5*]. Within the immigrant class, the longer a male had been a member of the group, the less

frequently females avoided him over long distances; this correlation was significant [*6*]. Collins (1981) also found that females avoided two newcomers more often than they avoided any other males.

4. When an infant expresses distress and a male is nearby, not only the mother but also her relatives and male Friends are likely to rush to the scene, pant-grunting and threatening the male, even when the infant's distress was not evoked by the male. In extreme cases, a whole group of females and juveniles will mob the male. Although no systematic data are available, in EC the likelihood that infant distress would evoke defensive responses seemed to depend on the identity of the male or males near the infant: Recent immigrants were most vulnerable to such responses, and males who were Friends of the mother were least vulnerable. I have seen male Friends swat annoying infants away, evoking prolonged screaming in the infants, and no one paid any attention, including the mother. On other occasions, I have seen a recent immigrant mobbed simply because he was within a few meters of an infant who began to whimper softly. Collins (1981) and Strum (1984) also reported that most mobbings were directed toward males who had recently moved into the group.

Taken together, these four sets of observations indicate that baboons tend to adopt a chronically suspicious attitude toward males, and especially toward newcomers, in situations involving infant distress. Males, in turn, show responses that appear designed to reduce their vulnerability to defensive mobbing by infant protectors.

1. Upon being approached by an infant, males sometimes exhibit submissive gestures (personal observation) and avoidance (Stein, 1981). These responses were most commonly seen in recent immigrants, and they were never observed in males who were Friends of the mother.

2. Whenever an infant expresses even mild distress (e.g., whimpering or gecking), most nearby baboons (with the exception of other infants) tend to look at the infant and grunt rhythmically until the infant's distress ceases (Ransom, 1971). Gilmore (personal communication cited in Stein, 1981) found that 90% of the distress vocalizations of black infants elicited grunts in other baboons. Similar grunt choruses are sometimes evoked simply by the proximity of a young infant who is off its mother's body. These grunts sound similar to grunts used during friendly approaches (Gilmore, 1980). Recently transferred males seemed to be particularly loud and active participants, and I had the impression that they were attempting to communicate to other nearby baboons their positive attitudes toward the infant. An alterna-

tive hypothesis, proposed by Stein (1981), is that a male grunts in order to familiarize the infant with his voice so that it will be more likely to respond positively to the male's grunts on future occasions when he is trying to carry the infant during an interaction with another male (male–infant carrying interactions are discussed in Chapter 8). This suggestion may apply to males who are in the process of developing a close bond with the infant, but it seems less applicable to recent immigrants, who hardly ever carry infants in interactions with other males (see Chapter 8).

The intense alarm shown by mothers when near recent immigrants, the fearful responses of infants to these same males, the suspicious attitude of troop residents toward newcomers who are near infants, and the attempts by males in general and newcomers in particular to avoid being implicated in infant distress—this diverse array of responses provides strong circumstantial evidence that male infanticide, particularly by recent immigrants, has been a potent selective force operating on baboon social behavior. If this hypothesis is correct, then protection of infants from infanticide could be a major benefit to females of friendships with males. Although there is little direct evidence of such protection, perhaps in part because male attacks on infants are so rarely observed, male baboons do form long-lasting, affiliative relationships with particular infants. These relationships and the ways they may benefit the mother and infant are reviewed below.

MALE–INFANT RELATIONSHIPS

Previous Research

In all of the major early studies of savannah baboons, observers noted that lactating females with young infants tended to associate with adult males (Hall, 1963; DeVore, 1963; Rowell *et al.*, 1968; Saayman, 1971a). Males were also seen greeting, holding, and carrying infants and protecting them from other troop members and predators. DeVore's account of male–infant relations was the most detailed of these early studies:

> *As soon as her infant is born, the mother moves to the heart of the troop to be near the oldest and most dominant males. . . . Regardless of her former position in the female dominance hierarchy, the mother is now virtually immune from threat by other troop members because she enjoys the complete protection of the dominant males. The males themselves are intent upon*

protecting the new infant, a protection which necessarily includes the infant's mother (1963, p. 312).

DeVore (1963) gave several examples of male care and protection of infants: "Adult males, even young adult males, allow black infants to crawl all over them with impunity" (p. 318); adult males "viciously attack any human who comes between an infant and the troop" (p. 315); during play with juveniles, "any sign of fear or frustration by the black infant causes an adult male to stare toward the play group, sometimes grunting softly, and the offending juvenile releases the infant immediately" (p. 319).

DeVore noted that "the degree of interest in infants shown by adult males varies considerably" (p. 314). Ransom and Ransom (1971) were the first to show that some of this variability could be explained by the male's relationship with the mother. In many cases, close male–infant relationships "derived from the pair bond of adults" and represented "a widening of the attachment between the male and the female to include the female's infant at birth" (p. 183). The authors' descriptive accounts illustrate the extreme closeness of some male–infant bonds: "Harry fulfilled a number of paternal functions for Myrna's offspring when they were out of contact with the mother. These included 'baby-sitting' (i.e., remaining with the infant, grooming and sometimes carrying it, and providing frequent, reassuring contact), and passive and active protection" (p. 184).

Cheney (1977b) and Seyfarth (1978b) provided quantitative evidence that males who had special relationships with a female also developed affiliative relationships with that female's offspring. Packer (1980) distinguished three criteria of male care of infants among olive baboons at Gombe: (1) percentage of time males spent within 5 m of an infant, used as a measure of protection of the infant against predators; (2) frequency with which males defended infants against other baboons; (3) frequency with which adult males groomed infants. All three types of care were significantly more common between infants and males who resided in the troop at the same time the infant was conceived than between infants and males who had recently transferred into the troop. Busse and Hamilton (1981) reported a similar result among chacma baboons: Males who were both present in the troop and active consorts around the time of the infant's conception were the males most likely to protect infants against other males in the troop. However, their measure of protection, the carrying of infants during interactions with other males, is a controversial one, and this aspect of male–infant relations is discussed in detail in Chapter 8.

Altmann (1980) considered more routine ways in which males might benefit infants and their mothers. Among Amboseli baboons, mothers with young infants experienced high rates of potentially stressful interactions with both male and female baboons, including approaches that evoked fear or distress in the mother, supplants, and attempts to pull the infant off the mother. The presence of a male with whom the mother had a friendship seemed to inhibit the behavior of other baboons, resulting in a reduction in the rates of stressful interactions experienced by the mother. As in this study, these males tended to be ones who had associated with the mother prior to the infant's birth, and they were sometimes probable fathers of the infant (see Chapter 8). In the first few months of life, associates helped to provide a buffer between the mother–infant pair and other troop members. As infant independence from the mother increased, relationships with these males continued to be important: "Older infants rested against their male associates and obtained rides from them, ran to them in times of distress, and took greater liberties with these males than with others" (p. 115). Altmann's case history descriptions of mother–infant pairs contain several striking instances of male care and protection of an infant with whom he had a friendship, including an unsuccessful attempt to rescue an infant from a fatal attack by another troop male. Stein (1981, 1984), working with the same troop, later confirmed and expanded Altmann's findings on male–infant relationships.

The pattern of male–infant interaction in EC was similar to that reported for Amboseli baboons. Males who had friendships with females tended also to have friendships with the infants of those females. Data on male–infant relationships were derived from three sources: (1) interactions between males and infants during focal samples on mothers; (2) interactions between males and infants observed during ad lib sampling; and (3) data on spatial proximity and grooming between males and infants derived from N. Nicolson's study of EC infants.

Affiliative Interactions between Males and Infants

Table 6.4 lists 6 different types of affiliative interactions that occurred between EC males and infants. During focal samples on females, male Friends accounted for most (88%) of these affiliative interactions, even though they were represented in only a small fraction (about 12%) of the male–female dyads in the troop (Table 6.4). Ad lib records of male–infant affiliative interactions produced a similar result: 22 of the 27 (74%) interactions involved Friends.

The 6 types of male–infant affiliative interactions can be divided into two classes: those that involved prolonged physical contact and

Table 6.4. Frequency of Affiliative Interactions between Males and Infants: Friends and Non-Friends[a]

Type of interaction	Percentage involving Friends	N
No prolonged touching		
Male grunts in response to infant distress	76.9	13
Infant approaches male[b]	91.1	56
Male greets infant[c]	80.0	40
Total	85.3	109
Prolonged touching[d]		
Male holds infant ventral	96.4	28
Male carries infant	100.0	7
Male grooms infant	100.0	3
Total	97.4	38
All types combined	88.4	147

[a]Values were derived from focal samples of mothers with infants less than 1 year old.

[b]Infants sometimes showed intense ambivalence when approaching a male, alternately screaming, threatening, and presenting to him while looking around at other baboons (Packer, 1979b); such approaches were never directed at males who had friendships with the infant's mother. These "ambivalent approaches" are not included here.

[c]Male greets infant includes all occasions where the male briefly touched the infant while grunting and/or lip-smacking at the infant within 1 minute after an approach.

[d]Holding, carrying, and touching that occurred during an agonistic interaction between two males are not included (see Chapter 8).

those that did not (Table 6.4). Only 1 of the 17 instances of affiliative interaction between Non-Friend males and infants involved prolonged physical contact. Additional information on male–infant affiliative interactions is available from Nicolson's focal samples of EC infants. Nicolson (1982) saw males groom infants 9 times and infants groom males 15 times. All grooming dyads involved males who were Friends of the infant's mother. Infants were observed riding on a male's back or ventrum 27 times, and all but four involved male Friends of the mother (the four exceptions involved Homer, a male thought to have been born in EC who apparently had a habit of carrying infants). These results, combined with those in Table 6.4, show that infants were held, carried, and groomed almost exclusively by Friends of the mother.

The intimate tone of interactions between Friends and infants is striking to the observer but difficult to describe. Several examples are included below to help convey what these relationships were like. The first example illustrates the relaxed, affectionate quality of the majority of interactions between male Friends and infants.

Leda is feeding with her infant clinging to her ventrum. Plato, her Friend, approaches, sits right next to her, and, lip-smacking and grunting, pulls the infant Laocoon off Leda and gathers him to his chest. Leda ignores Plato. Laocoon clings to Plato for a few seconds and then plays quietly in his lap. Plato becomes sleepy and his eyelids droop. Laocoon leaps up toward Plato's face and grasps the hair by his cheek as he falls back down, jerking Plato's head. Plato, without opening his eyes, grunts and curls one hand around Laocoon's body (focal sample on Leda, 21 October 1978, infant aged 23 weeks).

Despite the affectionate nature of their interactions with infants, male Friends sometimes upset infants inadvertently. The next example illustrates the kind of efforts males made to reassure their small associates when this occurred.

Boz is feeding about 2 m away from his Friend, Iolanthe, and her infant, Iona. Iona toddles toward Boz and leaps on to his back from behind. Boz, startled, jerks his body, scaring Iona. She gecks, and he sits down, pulling her around into his lap. He peers at her, grunting and lip-smacking, while touching her, and she soon stops vocalizing (focal sample on Iolanthe, 10 October 1977, infant aged 10 weeks).

Male Friends were generally extremely solicitous of young infants in distress, as illustrated by the next two examples.

Psyche and her Friend, Handel, are feeding near one another and Psyche's infant Penelope sits between them. A juvenile charges past Penelope and she screams. Psyche and Handel begin to move toward Penelope at the same time, and Handel reaches her first. He pulls her ventral, lip-smacking. Psyche resumes feeding; Penelope remains clinging to Handel for several seconds (focal sample on Psyche, 27 May 1978, infant aged 4 weeks).

Olympia is grooming her Friend, Achilles. Octavia, Olympia's infant, plays by herself a few feet away. She tries to climb a sandy mound and, almost to the top, she slips and slides rapidly to the bottom. Achilles, who was watching her antics, gives reassuring grunts as soon as Octavia slips (focal sample on Olympia, 16 November 1978, infant aged 10 weeks).

This last example is particularly interesting because Octavia showed no signs of distress. Achilles behaved as if he thought that slipping on the sand was an experience that deserved the same sort of reassurance he would normally give when the infant of a female Friend screamed or gecked in distress.

The following example, describing two events that occurred just seconds apart, illustrates the radical differences in the mother's responses to Non-Friend males and Friends when they attempted to interact with her infant.

Phoebe is grooming herself, and her infant Phyllis sits next to her. Hector, a Non-Friend resting about 2 m away, shifts closer to Phoebe and grunts at Phyllis. Phoebe flinches and pulls Phyllis toward her, away from Hector. She then turns her back on Hector, placing her body between him and Phyllis. Phoebe continues to groom herself, and Cyclops, a Friend, approaches, grunts, and touches Phyllis gently. Phoebe stops grooming and Cyclops starts to groom her. Phyllis climbs on Cyclops and hangs from his upraised arm, interfering with his grooming of Phoebe. Cyclops gently shakes his arm and Phyllis drops to the ground with a thud and a small whimper. Phoebe, whose eyes are now closed, shows no response. Cyclops resumes grooming and Phyllis soon climbs back onto his body (focal sample on Phoebe, 22 August 1978, infant aged 16 weeks).

Male Protection of Infants

The effects of the attachment that develops between a male and the infants of his female Friends are particularly obvious when the infant is endangered. At such times, both the mother and her Friends will run to the infant, and the infant, in turn, may run toward a Friend if he is closer than the mother. There are three contexts in which the proximity of a male willing to protect it might be of crucial importance to an infant: (1) when the infant is vulnerable to capture by a predator; (2) when the infant is vulnerable to harassment by another baboon; and (3) when the infant is vulnerable to male infanticide. Each of these contexts is considered in turn, below.

Predation. With the exception of the one attack by dogs mentioned in Chapter 2, no attempted predation of baboons was observed during the study period. However, on two occasions I inadvertently frightened an infant, and the responses of the males to me may indicate the types of responses one would observe to a true predator. In both cases, I walked very close to an infant, and it screamed. Both times, a Friend of the infant's mother charged me from over 30 m away, pant-grunting with mouth agape (in the second instance, the infant's mother charged me as well). Several other males just as close to the infant who did not have a friendship with the mother simply sat and watched. I turned and slowly moved away from the male, hoping to communicate disinterest in the infant. In both cases, the male's charge stopped less than a meter away from me, although he continued to threaten me for several minutes. Adult male baboons have been seen to effectively ward off leopards and lions in Masai Mara (R. Sapolsky, personal communication), so it is clear that they are capable of such active defense.

Ransom (1981) described a close relationship between Ringo, a male baboon at Gombe, and Lamb, an infant who was often neglected by her mother. Ringo was particularly solicitous of Lamb when chimpanzees were nearby (chimpanzees at Gombe prey on infant baboons). Ransom saw the chimpanzees make several attempts to capture Lamb, but each time they were "thwarted by Ringo's attentions" (p. 229).

Stein (1981, p. 179) provides a striking example of the differential protection provided by a male who has a close relationship with an infant:

> *During one alarm that I observed in response to eight lions, eight-month-old Gina ran to and jumped on the nearest adult male, Wart, with whom she had no close relationship . . . . Wart deliberately knocked her off and ran on. A few seconds later, Even, the adult male . . . with whom she had the closest social relationship, approached her, palmed her, and carried her away.*

Harassment. Male Friends also protect infants against other baboons. Altmann (1980) found that the rate of infant-pulling and handling during the infant's first month was usually lower when a close male associate of the mother was nearby. In the same troop of yellow baboons, Stein (1981) found that rates of harassment of infants by other troop members were significantly lower when a "preferred" male was within five feet of the infant than when he was not (preferred males were ones who had a special relationship with the infant; as in EC, these relationships generally developed as a result of a close relationship between the male and the infant's mother). The reduced rate of harassments was probably related to the fact that preferred males "punished other animals who harassed infants"; nonpreferred males did not defend infants in this way (Stein, 1981, p. 195).

Infanticide. Altmann (1980) describes an unsuccessful attempt by a male to rescue the infant of one of his female affiliates from a fatal attack by another male. At Moremi, young infant chacma baboons were twice seen clinging to resident males near a recent immigrant who had just attacked another infant. In two cases, one in EC and one at Gombe (Collins *et al.*, 1984), observers heard males fighting just before the attack on the infant. It is possible that these fights occurred when one male tried to protect the infant from an attack by another male. During most of the infanticidal attacks witnessed at Gombe and Moremi, however, males were not observed trying to protect the infants, although it is not clear that other males were actually present during the attacks; in several cases, females did try to interfere (Collins *et al.*, 1984). The fact that males were not seen to defend infants in these cases does not necessarily imply that male protection against infanticidal attacks is rare or unimportant. There is no way to know how often an infanticidal male refrains from attacking an infant because of the protective presence of another male; we see only those cases in which for one reason or another male protection fails to inhibit an attack (Packer, 1980). The more common male protection of affiliated infants is, the more likely it is that potentially infanticidal males will learn not to attack infants when a protective male is nearby. It is possible that most of the observed infantcidal attacks succeeded precisely because there were no males present willing to protect the infant, and that such attacks would be much more common if male protection did not occur.

Relationships between Older Infants and Adult Males

In their first 2 or 3 months of life, infants are rarely found far from their mothers, and proximity to adult males is almost completely determined by the mother's behavior and the behavior of the males. As they grow older and more independent, infants spend increasing amounts of time away from their mothers and begin to control their associations with other troop members. A striking aspect of the behavior of older infants is their tendency to associate with particular adult males when they are far from their mothers (Altmann, 1980; Nicolson, 1982). These associations were documented in EC by Nicolson, who conducted focal sample of infants from April 1978 to July 1979. At the start of each 1/2 hour focal infant sample, she noted the identities of all adult and subadult males near (within 5 m) the infant. There were 265 samples in which at least one male was near the infant (the mean number of males near the infant in these samples was 1.08, indicating that infants were rarely found near more than one male at a time). Nicolson then calculated for 20 infants the percentage of occasions each male in the troop was near the infant out of those occasions when any male was near it for two conditions: mother-present (mother within 5 m of the infant) and mother-absent (mother further than 5 m from the infant).

The results (Appendix XIV) show that when the mother was absent, most infants (15 of 20) were found primarily near one male; 5 infants associated with two males. Of these 25 male associates, 23 were Friends with the mother during the infant's gestation and first few months of life. Data on the mothers' proximity to the males who associated with their infants showed that infants, even more than mothers, tended to restrict their associations with males to one particular male (or, in a few cases, two males) [7]. These results suggest that infant–male associations were somewhat independent of the mother's associations. Independence between mother and infant associations is further indicated by the fact that the majority of infants over 16 weeks of age was more likely to be found within 2 m of a male when they were over 5 m away from the mother than when the mother was present (Nicolson, 1982).

These results suggest that, for older infants, adult males may help to fill a gap created by the weakening of the earlier, close relationship with the mother (Nicolson, 1982). This substitution of a male Friend for the mother is particularly striking in two situations: (1) when the infant is rejected by the mother unusually early and (2) when the mother dies. Nicolson (1982, p. 98) notes that "all four infants who were weaned from riding before one year of age developed strong relationships with one or two males." For example, Chloe, Clea's daughter, began a close association with Cyclops, Clea's Friend, after Clea abruptly prohibited Chloe from riding at 7 months of age. From 8–14 months of age, Chloe spent almost 25% of her time within 2 m of Cyclops—an amount of time roughly equal to that which other infants her age were spending near their mothers.

The relationship between Aurora's infant, Uranus, and her male Friend, Aristotle, provides another example. Aurora, a primiparous female, neglected Uranus from the moment of his birth until his death 7 months later. She often refused to carry him or let him nurse, and his cries of distress were a pervasive part of EC life for months. Aristotle, a subadult, probably natal male, virtually adopted Uranus, carrying him frequently, grooming him, and protecting him from the attentions of other troop members. Once when the troop ran away from an unfamiliar person, I saw Aristotle make a detour to rescue Uranus who was playing in a tree; Aurora had fled with the others without a backward glance. It is doubtful that Uranus would have survived his first few weeks without Aristotle's aid. (As far as we know, Aristotle was not related to Uranus.)

The existence of a close relationship with an adult male may prove crucial to an infant's welfare if its mother dies. Two EC infants who lost their mothers when they were between 1 and 2 years old received special attention from males who had been Friends with the infant's mother.[5] These males tolerated close following by the infants, groomed them, and defended them against other troop members. Both infants survived to at least 4 years of age. DeVore (1963) reported a similar relationship in a troop of olive baboons in Nairobi Park. An orphaned infant (aged 6 months to 1 year) who became ill was the "constant companion" (p. 323) of an adult male. The male protected her from other troop members and allowed her to feed near him. Thus, a special relationship with an adult male may increase an infant's chances of survival when it is neglected by or loses its mother.

Relationships with adult males probably also benefit older infants with more normal histories. Stein (1981) has provided a careful evaluation of these benefits. The two most important benefits appear to be continued protection from harassment, as described above for younger infants, and tolerance of infants feeding nearby. Because adult males are dominant to individuals of all other age/sex classes, they tend to occupy the best feeding sites, and so infants may derive significant benefits from feeding close to their male associates. Proximity to males is probably particularly advantageous when it allows infants to eat foods they cannot obtain on their own (such as scraps of meat or corms, which they are not strong enough to dig up). Stein (1981, p. 212) reports that male tolerance of infants feeding nearby "is overwhelmingly associated with infant–adult male dyads having preferred relationships."

Cofeeding between males and the infants of their female Friends was also observed in EC. Nicolson (1982, p. 95) describes an "antiphonal 'grunt duet' " that occurred between male associates and infants feeding near them (these duets were never heard between mothers and infants and rarely between infants and any partners other than male Friends): "In feeding situations, the grunt serves to maintain contact between the participants without interrupting visual attention to food selection and processing. The reciprocal nature of grunt duets suggests that adult male duetters were tolerant of co-feeding infants and may even have encouraged their presence."

[5] The mothers conceived before the study began, so there was no information about likely paternity for these orphans.

Long-Term Persistence of Male–Infant Relationships

The special relationships that developed between EC infants and their mothers' male Friends often persisted at least through the infants' third or fourth year of life (Johnson, 1984). This also appears to be the case among the yellow baboons of Amboseli (Stein, 1981). These long-term bonds are characterized by continued spatial proximity, grooming, and defense by the male. Thus, the friendships a female has made by the time she gives birth will sometimes continue to benefit her offspring years later as they approach the beginning of their own reproductive careers.

Effect of Female Reproductive Condition on Proximity to Male Friends

The results presented above suggest that male Friends provided important benefits to females with infants (e.g., reducing harassment by other females and juveniles) and to the infants themselves (e.g., protection). If this is the case, then we might expect mothers with infants to spend more time near their male Friends than pregnant females do. This hypothesis is supported by data on proximity of male Friends before and after the infant's birth (Table 6.5). During the first 16 weeks of the infant's life, females spent significantly more time near (within 15 m) of their male Friends (relative to time spent near other males) than they did during pregnancy [8]. (Note that this result cannot be explained by the lower sociability of pregnant females, since a relative rather than an absolute measure of proximity to Friends was employed; see Table 6.5). The relative amount of time spent near Friends declined slightly and then increased slightly over the next two 16-week blocks, but these values were not significantly different from one another or from the value for 0–16 weeks. I did not have enough data for females with infants older than 48 weeks to determine whether the time females spent near male Friends declined as the infants achieved independence. Even if the data were available, the results would be difficult to interpret since many females resume sexual cycles near this time, adding a confounding variable to the comparison.

Was the fact that females spent relatively more time near Friends after birth than before the result of changes in the behavior of females or of their male Friends? This is an important question, but, unfortunately, only five female–Friend pairs had large enough samples of approaches and leaves to determine responsibility for maintaining proximity during pregnancy. For four of these five, the female was *more* responsible for maintaining proximity after birth than before,

Table 6.5. Relative Proximity of Male Friends as a Function of Female Reproductive Condition[a]

Female reproductive condition	Median relative proximity score	Number of females	Number of Friend dyads
Pregnant	2.6	15	33
Infant			
0–16 weeks	3.8	26	53
17–32 weeks	3.3	20	38
33–48 weeks	3.5	8	16

[a]Because the amount of time females spent near others varied with the time of year and between females, I used a relative rather than absolute measure of proximity to Friends; this measure takes seasonal effects and individual variability into account. For each female for each reproductive condition, I calculated the median percentage of time she spent within 15 m of *all* males based on focal female instantaneous samples. The Friend's relative proximity score was the ratio of the percentage of time he spent within 15 m of the female divided by the median score for all males. Thus, a relative proximity score of 1 indicates that the female was near her Friend as often as she was near the male with the median proximity score, and a relative proximity score of 3 indicates that she was near her Friend three times as often as she was near the median male. The values shown in the first column are the median values for all female–Friend pairs for whom adequate data were available (i.e., at least 10 focal samples per female per reproductive condition).

indicating an increased desire to be near the male Friend. This result is consistent with the hypothesis that the mother–infant pair receives important benefits from the Friend's proximity.

SUMMARY

Friends were responsible for nearly all aid that females and their offspring received from males during agonistic interactions with other troop members. This aid benefits the females and juveniles (1) by reducing disruption of their activities and other forms of harassment (when the aggressors are other females and juveniles) and (2) by reducing rates of infliction of wounds (when the aggressors are subadult or adult males). However, friendship sometimes results in costs to females as well as benefits. These costs include increased vulnerability to aggression by males who use the female as a means of manipulating or challenging her male Friend and to mock attacks and use as an agonistic buffer by Friends themselves. Analysis of the complex relationships between males, between females and Non-Friend males, and between females and Friends, suggests that forming a friendship with one or two particular males may be the best response a female can make to the ever-present danger of aggression from other troop members.

Males that have recently transferred into a group have been observed killing infant baboons, and protection from infanticidal attacks may be another important benefit that females derive from friendships with males. This protection is most likely to come from male Friends of the mother, who were much more likely than other males to develop affiliative relationships with infants. Special relationships between males and infants were manifested routinely by proximity between the male and infant and by friendly interactions (e.g., greeting, holding, carrying, and grooming of the infant by the male). When away from their mothers, older infants tended to associate with one or, less often, two particular males, and these males were nearly always Friends of the infant's mother while she was pregnant and during the infant's first few months of life. Friendships with males may benefit infants in several ways, including protection from harassment by other troop members, protection of infants from predators and from other baboons, increased foraging efficiency, and the potential for alternative caregiving when the mother neglects the infant or dies. In some instances, these relationships—and presumably some of the benefits they provide—persist at least until the offspring's third or fourth year of life. The fact that females spent relatively more time near male Friends after giving birth supports the hypothesis that the development of male–infant relationships is one of the most important benefits that females derive from friendship with males.

These results suggest that when a female establishes a friendship with a male, she and her immature offspring acquire a long-term ally in the troop—an ally who, because of his larger size and superior fighting ability, may make a significant contribution to the survival and reproduction of the female and her offspring. If these conclusions are correct, they provide a functional explanation for the females' participation in friendships with males. I will consider shortly the benefits males may derive from friendship with females. In order to evaluate the benefits to males, however, it is first necessary to explore in some detail males' relationships with one another and the ways in which they compete for access to mates. This is the subject of the following chapter.

NOTES ON STATISTICS

[*1*] Comparison of the observed versus expected frequencies of male aggression toward females by Friends and Non-Friends. Expected

frequencies were determined by multiplying the total number of observed male–female aggressive episodes by the proportion of all male–female dyads that were Friends (12%) and Non-Friends (88%). $\chi^2 = 7.95$, $d.f. = 1$, $N = 93$, $p < .01$.

[*2*] Comparison of the observed versus expected frequencies of male aggression toward females by Friends and Non-Friends. Expected frequencies were based on the proportion of all male–female dyads that were Friends times x, where x = the average amount of time spent within 15 m of females by Friends divided by the average amount of time spent within 15 m of females by Non-Friends. The value of x (3.5) was based on focal samples of lactating females. $\chi^2 = 9.26$, $d.f. = 1$, $N = 93$, $p < .01$.

[*3*] Comparison of the observed versus expected frequencies of chases, threats, and attacks by Friends and Non-Friends. Expected frequencies based on overall participation by Friends and Non-Friends in aggression toward females. $\chi^2 = 1.69$, $d.f. = 2$, $N = 93$, n.s.

[*4*] Comparison of the observed versus expected frequencies of severe aggression and mild aggression when the aggression was "unprovoked." Expected frequencies were based on the proportion of all male—female aggression that involved severe aggression. $\chi^2 = 2.97$, $d.f. = 1$, $N = 22$, $.05 < p < .10$.

[*5*] Comparison of proportion of movements to within 5–15 m that elicited female long-distance avoids by short-term residents and newcomers versus long-term residents. Fisher exact probability test, $N = 30$, $p = .04$.

[*6*] Correlation between length of time a male had been a troop member and proportion of movements that elicited long-distance avoids by females. Spearman rank correlation, $r_s = -.83$, $N = 6$, $p < .05$ (one-tailed).

[*7*] Comparison of the number of male–infant dyads in which the male accounted for a greater percentage of the infant's time near males than of the mother's time near males versus the number of dyads in which the reverse was true. Sign test, $x = 16$, $N = 23$, $.05 < p < .10$.

[*8*] Comparison of relative proximity scores (see Table 6.5 for definition) of Friends during pregnancy and during the infant's first 16 weeks of life. Only those pairs for whom adequate data were available for *both* reproductive conditions were included ($N = 15$ females, 33 pairs of Friends). Mann–Whitney U test: $U = 406$, $z = 1.78$, $n_1 = 33$, $n_2 = 33$, $p < .05$ (one-tailed).

7 MALE–MALE COMPETITION FOR MATES

Two adult male allies jointly threaten a third male (not shown). By forming stable coalitions, older, less dominant males are able to take estrous females away from younger, high-ranking opponents. (Photo by Jim Moore, Anthrophoto)

INTRODUCTION

Evolutionary theory employs two principles to explain who mates with whom: competition between members of one sex for access to members of the other sex (usually male–male competition for females), and preferences by members of one sex for particular members of the other sex (usually female choice of some males over others) (Darwin, 1871; Fisher, 1930; Bateman, 1948; Trivers, 1972). In baboons and in many other species, both male–male competition and female choice appear to be important determinants of mate selection; the two are usually so closely related that neither can be appreciated fully without considering the other. This chapter focuses on male–male competition for mates. It provides a context for interpreting female choice and male investment in offspring, topics considered in the following chapter.

I begin with an analysis of male consort activity in EC, because this is the best available measure of the outcome of competition for mates. The results show that some males spent much more time in consortship with potentially fertile females than did others. In order to understand why this was so, I investigate the tactics males used to acquire consort partners and how these tactics varied as a function of agonistic rank and residence status. I also consider psychological aspects of consort harassment. The findings indicate that consort success was strongly influenced by cooperative relationships with other males and by individual characteristics that allowed a male to acquire and maintain consortships without recourse to risky fights.

MALE CONSORT ACTIVITY, AGONISTIC RANK, AND AGE/RESIDENCE STATUS

Consort Success among EC Males

Ad lib data on sexual consorts were used to estimate mating success among EC males. Throughout the study period, whenever a male was seen in consort with an estrous female, the identities of the participants and the time of the observation were noted. On some days, observation conditions allowed nearly continuous monitoring of consort activities. On others, monitoring was possible only at irregular intervals due to constraints imposed by the focal sample protocol. However, since consort pairs rarely travelled outside the main body of the troop, intermittent monitoring is unlikely to have biased the results.

Male consort scores were based on the identities of consorting males for days D-7 through D-1 of the female cycle since conception is most likely to occur on those days (Hendrickx and Kraemer, 1969; Wildt *et al.*, 1977; Shaikh *et al.*, 1982), and since those are the days when most consort activity occurs (Hausfater, 1975; Rasmussen, 1980) (see Chapter 2 for a general description of female cycles). Each day was divided into a morning and an afternoon period. For scoring purposes, each period was considered worth 1 point, for a maximum of 14 points per female cycle (due to gaps in observations, on average each female cycle contributed 6 points to the total score). If only one male was observed in consort with a given female during a given period, he received 1 point; if two males were seen consorting each received .5 point, and so on. There were no cases in which more than three males were seen in consort with one female during a period. The results described below are based on 22 estrous cycles that resulted in conception during the first year of the study when male troop membership was constant.

Figure 7.1, which shows total consort scores for different males, indicates great differences in consort activity between males. The most successful male, Cyclops, accounts for 28% of the total score; the two most successful males together account for 43% of the total; and the five most successful males, only one-third of all the males ever seen in consort, account for nearly three-quarters (74%) of the total score. Such large individual differences in consort activity are typical of savannah baboon males (Hall, 1962; DeVore, 1965; Saayman, 1971a; Ransom, 1971, 1981; Hausfater, 1975; Seyfarth, 1978a; Popp, 1978; Packer, 1979a,b; Harding, 1980; Rasmussen, 1980; Busse, 1981; Collins, 1981; Manzolillo, 1982; Strum, 1982; Sapolsky, 1983).

Consort Selectivity

Before considering the effects of agonistic rank and age/residence status on consort success, two factors that influence male mating preferences will be discussed briefly.

The Probability That a Consortship Will Lead to Conception. As noted in Chapter 2, female baboons normally have several estrous cycles before they conceive, and during fertile cycles, conception is much more likley to occur during a 2- or 3-day period near the end of tumescence than earlier in the tumescent phase (Hendrickx and Kraemer, 1969; MacLennan and Wynn, 1971; Wildt *et al.*, 1977; Shaikh *et al.*, 1982). Copulation is therefore likely to result in conception during only a small fraction of the total time that a female shows sexual swellings and is sexually receptive. If males have any means of estimating those times when conception is most likely to occur, we would expect to see them mating selectively.

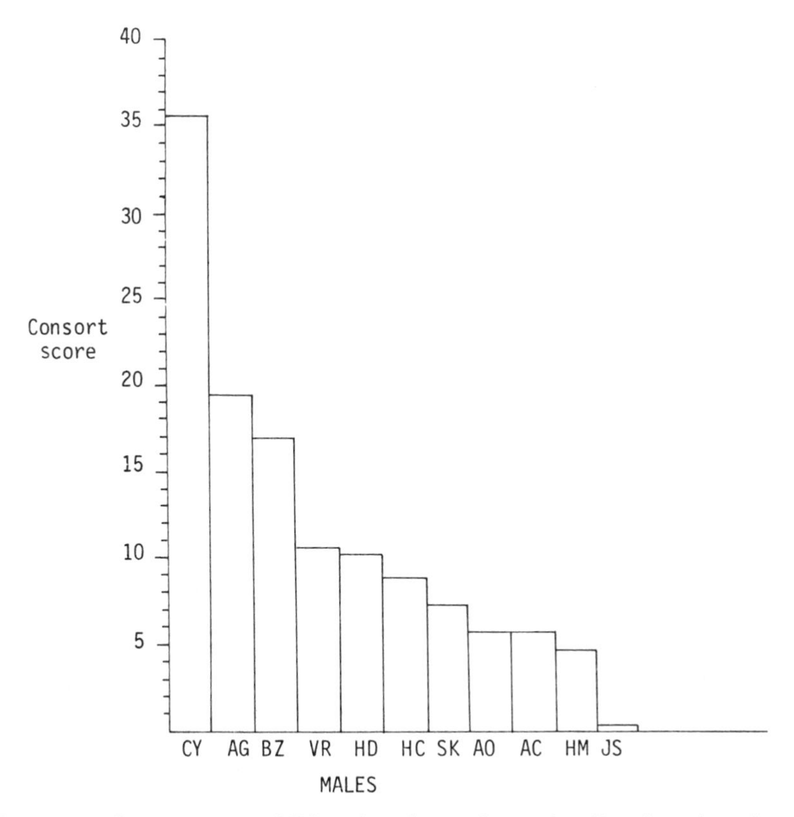

Figure 7.1. Consort scores of EC males: Conception cycles. Based on data from 22 conception cycles. See text for method of determining male consort scores. Seven males who were never seen in consortship during days D-7 through D-1 of conception cycles are not shown: adults AA, TN, and IA and subadults AS, HS, PL, and PX.

Cyclops, the most successful consorter in EC, appeared to discriminate between conception and nonconception cycles. He was responsible for 28% of the total consort score for conception cycles but for only 16% of the total score for nonconception cycles. This difference reflects a more general pattern: The distribution of consort scores was more skewed for conception cycles (Figure 7.1) than for nonconception cycles (Figure 7.2), although this difference was not significant [*1*].

Cyclops also appeared to discriminate days that were more likely to lead to conception *within* cycles (Figure 7.3). His consort activity was highest on day D-3 and surrounding days, the days when mating is most likely to lead to conception according to the most recent review of

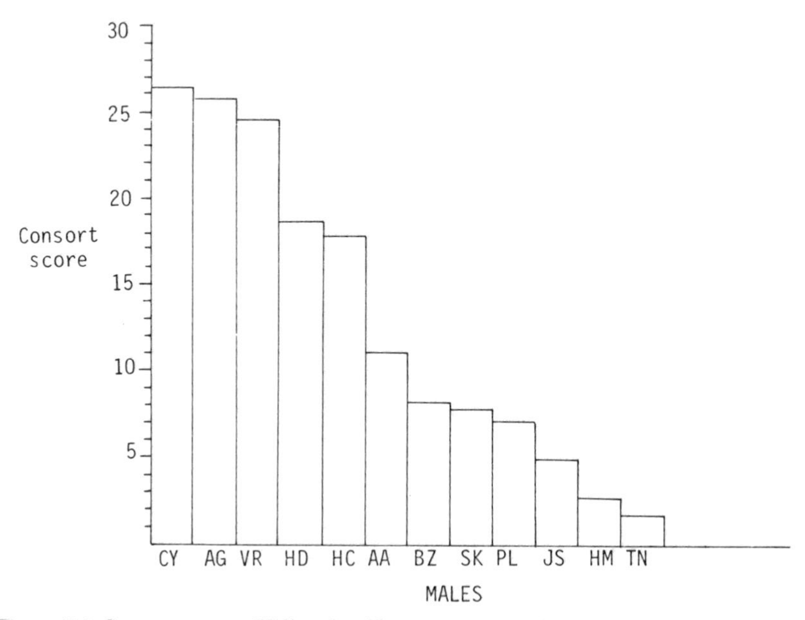

Figure 7.2. Consort scores of EC males: Nonconception cycles. Based on 28 nonconception cycles. See text for method of determining male consort scores. Six males who were never seen in consortship during days D-7 through D-1 of nonconception cycles are not shown: adults AC, AO, and IA and subadults AS, HS, and PX.

the laboratory evidence (Shaikh *et al.*, 1982, p. 450). Cyclops's apparent ability to estimate the timing of female cycles was so striking that I was often able to use his behavior to predict the day when a female's sexual swelling would begin to go down. I once saw him give up a consort partner and then quickly form another consortship with a different female. I later determined that he had abandoned the first female late on the afternoon of D-2 to establish a consortship with the second female late on D-4—the ideal strategy for maximizing his probability of fertilizing both females. Hausfater (1975) and Packer (1979b) also found that high-ranking males (in their troops these were the males with the highest consort scores) concentrated their consort activity around day D-3, but this pattern is not always evident (Seyfarth, 1978a; Rasmussen, 1980; Collins, 1981).

Relationships with Females. EC males also showed consort selectivity by contesting access to particular females and ignoring others, even when those females were undergoing conception cycles. Cyclops, for example, was never seen in consort with 9 of the 22 females whose

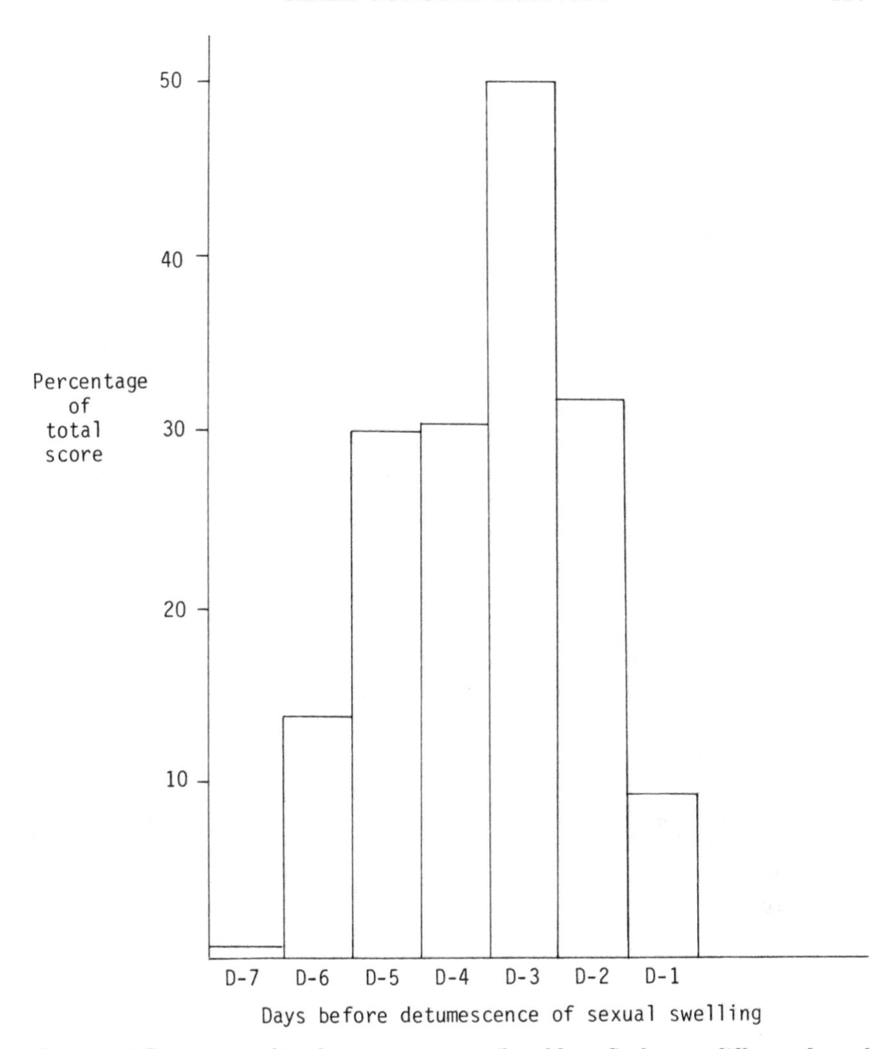

Figure 7.3. Percentage of total consort score attributable to Cyclops on different days of the female cycle. Based on 22 conception cycles. See text for method of determining consort scores.

conception cycles were analyzed above, and he showed no interest in these females, failing to follow or harass males who were in consort with them. Seven of these nine females consorted often with either Agamemnon, Boz, or Virgil—the three males who ranked two, three, and four on overall consort activity—and in each case, the consorting male *was* a Friend of the female and Cyclops was *not*. It thus appears that Cyclops avoided competing for those females who were most favored by his closest rivals. Several observers have reported that males sometimes avoid contesting access to females who have strong associations with other males outside the consort relationship (Ransom, 1971, 1981; Seyfarth, 1978a; Packer, 1979b; Collins, 1981). The relationship between friendship and consort activity is discussed further in the following chapter.

Agonistic Rank and Consort Activity

Previous investigators have determined agonistic rank among male baboons by examining the aggressive and submissive gestures exchanged during agonistic interactions (Hausfater, 1975; Popp, 1978; Strum, 1982), by analysis of approach/retreat interactions that do not involve aggressive or submissive gestures (Packer, 1979a,b) or by a combination of both methods (Sapolsky, 1983; Manzolillo, 1982).

Both methods were used to determine the winner of agonistic interactions among EC males. However, only 101 dyadic agonistic interactions between adult males were observed during 150 hours of male focal samples and ad lib observations combined.[1] Of these 101 cases, only 72 were decided agonistic bouts—bouts in which a clear-cut winner could be determined. In the remaining 29 cases, it was not possible to determine a winner either because one male ignored another's threats or because one male showed submission only after being threatened or chased by a coalition of two or more males. Thus, the classical dyadic agonistic encounter in which there is a clear-cut winner was rare in EC during this period.[2] Hall and DeVore (1965) also found that clear-cut dyadic agonistic encounters were rare in olive baboons.

[1] Dyadic agonistic interactions over access to meat and consort partners have been included.

[2] During the first few weeks of the study, before I began to record male approach–retreat interactions systematically, decided agonistic bouts were much more common. They were also common among EC males during a 3-week period in the summer of 1983 but rare during the rest of that summer (Smuts and Watanabe, in preparation). During the two periods when decided agonistic bouts were common, the troop was feeding almost exclusively on corms, a resource that is distributed in discrete patches and that involves a significant amount of preparation. The vast majority of decided agonistic bouts among males involved feeding supplants over corm sites. During other times, when the baboons ate grass or other more evenly distributed foods, feeding supplants among adult males were infrequent.

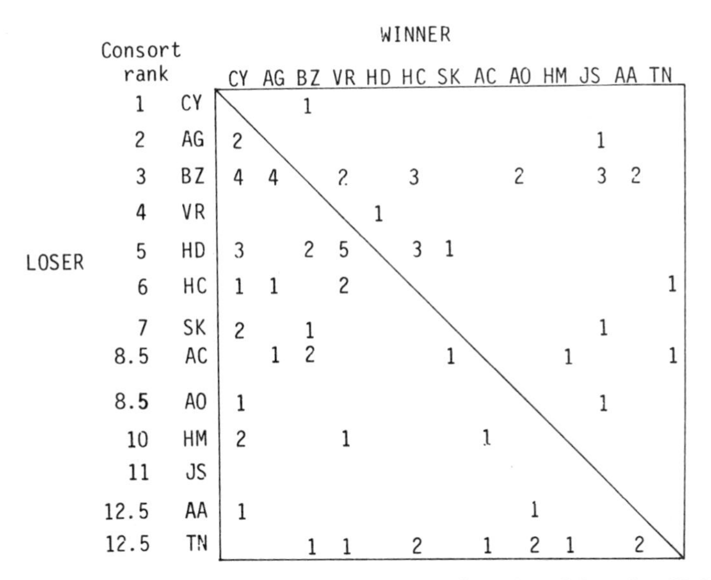

Consort rank	LOSER \ WINNER	CY	AG	BZ	VR	HD	HC	SK	AC	AO	HM	JS	AA	TN
1	CY			1										
2	AG	2										1		
3	BZ	4	4		2		3			2		3	2	
4	VR					1								
5	HD	3		2	5		3	1						
6	HC	1	1		2									1
7	SK	2		1								1		
8.5	AC		1	2				1			1			1
8.5	AO	1										1		
10	HM	2			1				1					
11	JS													
12.5	AA	1								1				
12.5	TN			1	1		2		1	2	1		2	

Figure 7.4. Outcomes of dyadic agonistic interactions between adult males: Males listed in order of consort rank. $N = 72$ interactions. The consort rank order is based on consort scores for 22 conception cycles (see Figure 7.1). Numbers to the left of the diagonal represent outcomes that were inconsistent with consort rank ($n = 48$ or 67%). One adult male, Ian, is not included because he was not observed participating in a decided agonistic encounter.

Despite the low frequency of decided agonistic bouts in EC, it may be of interest to compare the outcomes of such bouts with consort activity. Figure 7.4 shows a dominance matrix for the adult males in EC. The males are listed in order of *consort rank* in order to illustrate the relationship between consort activity and agonistic rank. The outcome in more than one-half (67%) of the agonistic bouts was inconsistent with consort rank. (When only the focal sample data from the last 3 months of the study period were used, a similar proportion of inconsistent outcomes emerged.)

It is possible to reorder the males in a way that minimizes the frequency of reversals, and the results are shown in Figure 7.5. A rank order based on this matrix is negatively, but not significantly, correlated with the rank order for consort success [2]. The validity of this result can be questioned, since the sample size on which the agonistic ranks are based is so small (Appleby, 1983). However, the absence of a positive relationship between agonistic rank and consort activity has been reported elsewhere. In the first detailed study of male–male

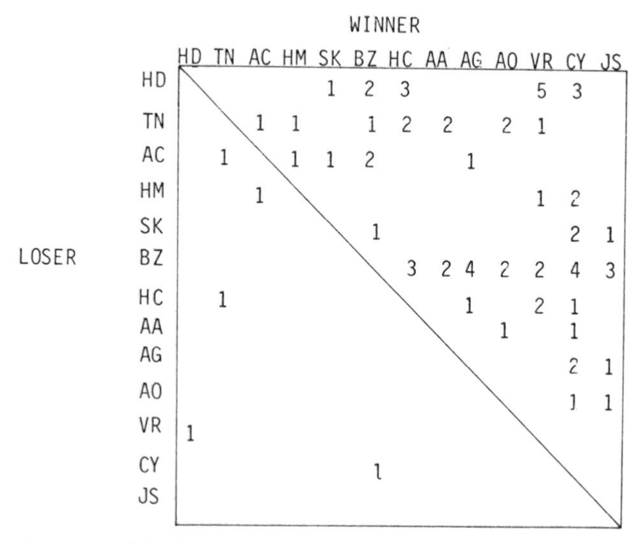

WINNER (columns) / LOSER (rows)

	HD	TN	AC	HM	SK	BZ	HC	AA	AG	AO	VR	CY	JS
HD					1	2	3				5	3	
TN			1	1		1	2	2		2	1		
AC		1		1	1	2			1				
HM			1								1	2	
SK						1						2	1
BZ							3	2	4	2	2	4	3
HC		1							1		2	1	
AA										1		1	
AG												2	1
AO												1	1
VR	1												
CY						1							
JS													

Figure 7.5. Outcomes of dyadic agonistic interactions between adult males: Males listed in the order that minimizes reversals to the left of the diagonal. This figure shows the same 72 outcomes included in Figure 7.4.

relationships in savannah baboons, DeVore (1965, p. 282) concluded that "the most effective breeders . . . may be subordinate as individuals to other males in the troop," and Saayman (1971a) reported similar results in chacma baboons. More recently, Strum (1982), working with the troop adjacent to EC, found no significant positive correlations and one significant negative correlation between agonistic rank and nine different measures of consort activity. Her agonistic rankings were based on a large sample of decided agonistic bouts recorded over several years of study. In the same troop at a later time, Manzolillo (1982) reported no significant correlations between agonistic rank and consort rank during any of nine consecutive 4-month periods of observation. Finally, during a study of male–male relationships among EC males in 1983, there was again no significant correlation between agonistic rank and consort rank (the results of the 1983 study are discussed further below). Thus, it is clear that in some baboon troops at some times, agonistic rank does not predict consort activity.

In contrast to these findings, a number of baboon studies have reported significant positive correlations between agonistic rank and mating activity (Hausfater, 1975; Seyfarth, 1978a; Popp, 1978; Packer, 1979a,b; Rasmussen, 1980; Busse and Hamilton, 1981; Collins, 1981;

Sapolsky, 1983). Possible reasons for these differences between troops are considered in a later section.

Relationship between Agonistic Rank and Age/Residence Status

In EC, young adult males generally had higher agonistic ranks than did older, long-term resident males; this was true for both natal young adults and young adults that had transferred from other troops (Table 7.1). Overall, young adults won 85% of their dyadic agonistic bouts with older, long-term resident males. Similar findings have been reported for the adjacent PHG troop (Strum, 1982; Manzolillo, 1982). In most baboon troops, as in EC, the male with the highest agonistic rank is a young adult who has recently transferred into the troop (Ransom, 1971, 1981; Rasmussen, 1980; Busse, 1981; Collins, 1981), although natal young adults also sometimes achieve alpha status (see data below for EC in 1983). These patterns result in a significant *negative* corre-

Table 7.1. Age/Residence Status, Agonistic Ranks, and Consort Success of EC Males: 1977–1978[a]

Male	Consort score	Consort rank	Agonistic rank	Mean consort score (median)
Older, long-term resident males:				18.6 (17)
AG	20	2	9	
BZ	17	3	6	
CY	36	1	12	
HC	9	6	7	
VR	11	4	11	
Young adult males:[b]				4.4 (5.5)
AA[t]	0	12	8	
AC[u]	6	8.5	3	
AO[n]	6	8.5	10	
HD[t]	10.5	5	1	
HM[n]	5	10	4	
IA[t]	0	12	—	
SK[t]	8	7	5	
TN[t]	0	12	2	
All males:				9.9 (8.0)

[a]See Table 2.4 for male weights.
[b]*t*, Transfer; *u*, troop of origin unknown; *n*, natal.

lation between age and agonistic rank among fully adult male baboons (Packer, 1979a; Popp, 1978; Rasmussen, 1980).

Age/Residence Status and Consort Activity

All five of the older, long-term resident males in EC had consort scores above the median, while this was true for only one of the young adult males (Table 7.1). Older residents had higher consort activity, on average, than young adult males, despite the fact that the younger males, in most cases, had higher ranks. The same pattern has been documented in the PHG troop by Strum (1982) and Manzolillo (1982). DeVore (1965) and Collins (1981) also observed long-term resident males who had higher consort success than expected based on their individual dominance ranks. Saayman (1971a) and Rasmussen (1980) do not discuss length of residency, but both provide examples of older males of low agonistic rank who had high consort success; similarly, Packer (1979b) reported a number of older males whose consort ranks were higher than their agonistic ranks.

Male Consort Activity, Agonistic Rank, and Age/Residence Status in 1983

As noted in Chapter 3, only limited data on male–male interactions were available from the 1977–1978 study (Study 1) because of my focus on anestrous females. However, the 3-month study conducted in the summer of 1983 (Study 2) provided systematic data on male–male interactions and, in particular, on competition over estrous females. For this reason, I have used observations from both studies in the following discussion of male competitive tactics.

Table 7.2 shows the males included in Study 2, along with their weights, age/residence status, agonistic ranks, and consort scores. (Note that only three of the adult males in EC in 1977, AA, BZ, and SK, were still in the troop at this time.) Male consort scores for Study 2 were determined in the same way as above, except that nonconception cycles were included, since only a few conceptions occurred during the study. As in Study 1, some males were much more successful consorters than others.

Male agonistic ranks were based on 175 dyadic agonistic bouts with decided outcomes (Figure 7.6). In 1983, *all* young adult males ranked above older, long-term resident males, and young adults won 97% of their dyadic agonistic interactions with older residents.

In contrast to Study 1, the two most successful consorters in 1983 were young adult, natal males. However, 75% of the older resident males scored above the median compared with only 38% of the young

Table 7.2. Age/Residence Status, Weights, Agonistic Ranks, and Consort Success of EC Males: 1983[a]

Male (weight kg)[b]	Consort score	Consort rank	Agonistic rank	Mean consort score (median)
Older, long-term resident males:				10.3 (12.0)
AA (30.5)	12	4.5	11	
BZ[c] (30.0)	4[c]	9.5	10	
SK (24.5)	12	4.5	12	
ZM (27.5)	13	3	9	
Young adult males:[d]				8.0 (5.5)
BA[n] (25.0)	2	12	3	
DT[n] (23.5)	18	1	1	
GR[u] (25.0)	5	8	8	
JP[t] (23.0)	6	7	6	
OR[n] (25.0)	10	6	4	
OV[n] (22.5)	16	2	2	
VL[t] —	3	11	7	
VU[n] (27.5)	4	9.5	5	
All males:				8.8 (8)

[a]Consort scores were based on both conception and nonconception cycles (N=14 cycles from 11 different females).
[b]Weight data were provided by R. Sapolsky, who captured EC males for a study of hormones and aggression. Weights for SK, GR, and VU were obtained in August 1982, 1 year before Study 2. Weights for all other males were obtained in August 1983.
[c]BZ's consort score was depressed as the result of an injury. See Footnote 3, this chapter.
[d]*n*, Natal; *u*, troop of unknown origin; *t*, transfer.

adults, and the mean and median consort scores of older residents were higher than those of young adults (Table 7.2).[3] Thus, despite their lower individual ranks, older residents had higher consort activity on average than younger adults during both studies.

In order to understand why older, long-term residents achieved high consort success despite their low agonistic ranks, we will now con-

[3] Part way through the study, one of the older resident males, BZ, suffered a deep gash on the palm of one hand (cause unknown). This injury eventually healed, but it severely interfered with his ability to compete for estrous females during the remainder of the study period. Based on BZ's behavior prior to the injury, it is likely that his normal consort activity was roughly equal to that of the other older resident males.

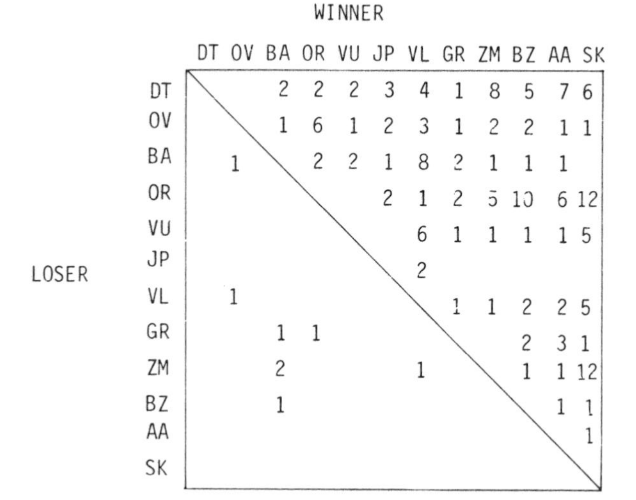

LOSER \ WINNER	DT	OV	BA	OR	VU	JP	VL	GR	ZM	BZ	AA	SK
DT			2	2	2	3	4	1	8	5	7	6
OV			1	6	1	2	3	1	2	2	1	1
BA		1		2	2	1	8	2	1	1	1	
OR						2	1	2	5	10	6	12
VU							6	1	1	1	1	5
JP							2					
VL		1						1	1	2	2	5
GR			1	1						2	3	1
ZM			2				1			1	1	12
BZ			1								1	1
AA												1
SK												

Figure 7.6. Outcomes of dyadic agonistic interactions between adult males during Study 2: Males listed in the order that minimizes reversals to the left of the diagonal. This matrix is based on 175 interactions observed between June and September 1983.

sider, in some detail, how males acquired and held on to their consort partners.

MALE COMPETITIVE TACTICS: DIFFERENT WAYS OF ACQUIRING CONSORT PARTNERS

Introduction

Male baboons compete, often intensely, over access to estrous females, and females usually change consort partners frequently, especially during peak estrus. For this reason, males who are able to win females away from other males and who, having established a consort relationship, are able to resist challenges by other males, will tend to have the highest consort activity and father the most offspring. In this section, I use data on successful consort "takeovers" to explore differences in the tactics males used to acquire consort partners from other males.

Before proceeding, it is important to note that I have used phrases like "acquire females from other males" and "consort takeovers" for the sake of convenience only, to refer to the relationship between males who are competing for opportunities to consort with the same female. Use of such terms is not meant to imply that females are passive objects of competition. In fact, as shown in the next chapter, female cooperation appears to be a critical factor in determining male consort success.

Frequency of Consort Turnovers

During peak estrus [the 7 days preceding rapid detumescence of the sexual swelling (Hausfater, 1975)], female baboons often change consort partners several times. Direct observations of consort turnovers in baboons are rare both because it is difficult to predict when they will occur and because they usually happen very quickly. For this reason, I used ad lib records of consorting pairs to estimate the frequency of consort turnovers in EC during Study 1.[4] Consort turnovers were assumed to have occurred when two (or more) different males were observed in consort with the same female on the same day. Turnovers were also assumed to have occurred when the female was with one male at the end of one day's observations and with another at the start of the following day's observations. The results (Table 7.3) indicate that during peak estrus, EC females experienced a mean of 4.8 consort turnovers involving, on average, four to five different males. (It is important to note that these estimates represent the *minimum* number of consort turnovers, since constant monitoring of estrous females was not possible.) Rasmussen (1980) reported very similar results (4.5 consort turnovers during peak estrus) in a troop of yellow baboons with approximately the same number of adult females and males as in EC.

Types of Consort Turnovers

During Study 2, we observed 21 consort turnovers during focal samples of consorting females (see Table 7.4).[5] Of the turnovers, 57% involved aggression between the previous consorting male and his successor, and 76% included aggression between the previous consorting male and at least one other male. However, most of this aggression involved threats only; the successor fought with the previous consorting male only once. Collins (1981) reported a similar proportion of

[4] Most consort turnovers involve an immediate shift from one partner to another (e.g., Hausfater, 1975). Occasionally, however, one consort relationship ends and the female remains alone for anywhere from a few minutes to several hours before another consortship is established. These types of consort turnovers are most common outside of peak estrus, when conception is unlikely to occur (Collins, 1981; personal observation). They are rare during peak estrus, the period used in this analysis.

[5] Consort turnovers that occurred outside peak estrus (days D-7 through D-1) and those that occurred during "resumption" cycles (Collins, 1981), the first estrous cycle after postpartum amenorrhea, have been excluded, since conception is unlikely to occur at these times (Scott, 1984).

Table 7.3. Mean Frequency of Consort Turnovers and Mean Number of Different Consort Partners during Peak Estrus[a]

Type of cycle	Mean number of consort turnovers (range)	Mean number of different consort partners (range)
Conception (N=30)	4.5 (1 – 6)	4.6 (1 – 6)
Nonconception (N=58)	5.1 (1 – 7)	4.4 (1 – 7)
All cycles (N=88)	4.8 (1 – 7)	4.5 (1 – 7)

[a]Following Rasmussen (1980), the mean number of consort turnovers was first determined for each day of the cycle (D-7 through D-1). These values were then averaged to estimate the maximum number of different partners per cycle. Note that the maximum number of different partners per cycle = the mean number of consort turnovers + 1. The values shown in the table are less than the maximum because a male sometimes consorted with a female more than once during a given cycle.

aggressive takeovers (55%) among yellow baboons during peak estrus of nonresumption cycles.[6]

Consorting males were subjected to two kinds of takeover attempts: coordinated harassment by two or more allied males and challenges by a single rival.

Coalitionary Takeovers. Table 7.4 shows that coalitionary challengers were usually older, long-term resident males of low agonistic rank, whereas the targets of these challengers were always young males of high rank (refer to Table 7.2 for individual ranks). In every case, all challengers ranked below the target. Young adult males occasionally joined older residents to challenge another young adult, but they never succeeded in acquiring the female. Each of the four older residents won the female at least twice, and agonistic rank did not seem to determine which of the allied males won the female. Only one coalitionary

[6] This value is based on data shown in Table 7.VIII, p. 281, Collins (1981). Strum (1982) reported much lower rates of aggression during consort turnovers in PHG troop. It is difficult to compare her results with those from EC and Collins' troop, because she did not indicate which cycles or which cycle days were included in the analysis. In Collins' troop, the highest proportion of aggressive turnovers occurred on days D-3 and D-2, when conception is most likely to occur. Aggressive turnovers were significantly less common outside of peak estrus and during the first cycle after postpartum amenorrhea, when conception is less likely to occur. Qualitative observations from EC were consistent with Collins' findings. It is possible that the lower rate of aggressive turnovers reported for PHG results, at least in part, from inclusion of turnovers occurring at times when competition over estrous females was reduced due to low probabilities of conception.

Table 7.4. Summary of Observed Consort Takeovers during Peak Estrus: Study 2[a]

Challenger(s)	Previous consort	New consort	Challenger(s) higher ranking than consort	Aggression
Coalitionary takeovers				
AA + 2 unidentified males	OV_y	AA	–	+
AA, SK, OR_y	OV_y	SK	–	+
AA,BZ,SK	DT_y	BZ	–	+
AA,BZ,SK	DT_y	BZ	–	+
BZ, ZM	DT_y	ZM	–	+
AA,BZ,SK,ZM	DT_y	AA	–	+
SK,ZM	DT_y	SK	–	+
AA,BZ, ZM,OR_y	BA_y	ZM	–	+
AA, ZM, VU_y	BA_y	AA	–	+
SK,ZM	DT_y	SK	–	+
AA,BZ,SK,ZM	DT_y	ZM	–	+
SK,ZM	BA_y	AA	–	+ (O)
SK, VU_y	DT_y	SK	–	+
Single takeovers				
VU_y	AA	VU_y	+	–
DT_y	BZ	DT_y	+	–
DT_y	BZ	AA	+	+ (O)
DT_y	AA	DT_y	+	+
GR_y	SK	GR_y	+	–
DT_y	ZM	SK	+	+ (O)
BZ	BA_y	BZ	–	–
None[b]	DT_y	SK		– (O)

[a]A challenger was defined as any male that attempted to take over a consortship through persistent following and/or harassment. Aggression was defined as any threat, chase, or fight between at least one of the challengers and the consorting male during the takeover. (O) indicates an opportunistic takeover, in which the new consort male was a male not directly involved in challenging the old consort. The subscript "y" indicates that the male was a young adult. Males without a subscript were older, long-term resident males.

[b]See text.

challenge resulted in an "opportunistic" takeover (Strum, 1982) by a male who had not challenged the previous consorter.

Single Takeovers. Single takeovers were less common than coalitionary takeovers. The roles of challenger and target were the reverse of those described above: Young adult males of high agonistic rank usually engaged in single challenges of older resident males of low agonistic rank, and the challenger ranked above his opponent in six of seven cases. Five of the eight single takeovers occurred without aggression. In four of the five, the consorting male moved away from the female after persistent following by the challenger, and in one case (the last listed in Table 7.4), a male formed a new consortship after the female ran away from her previous partner (this case is not included, below, when I refer to "single challenges"). In two of the three single takeovers that involved aggression between the challenger and the previous consorting male, an older resident male accomplished an opportunistic takeover while the challenger and target were exchanging threats or fighting. This suggests that aggressive challenges by high-ranking males were quite vulnerable to exploitation by lower-ranking bystanders.

Discussion. When the results from both types of takeovers in Table 7.4 are combined, we find that low-ranking, older resident males gained females much more often than did young adult, high-ranking males (81 versus 19%, respectively) and lost females far less often than did younger males (33 versus 67%, respectively). These findings are consistent with the higher consort scores of older resident males in EC.

The pattern of consort takeovers described above has also been reported for other troops. In the first detailed study of male baboons, DeVore stressed the importance of coalitions among particular males. DeVore (1962, 1965) and Hall and DeVore (1965) reported that by repeatedly forming coalitions with one another, these "central" males were able to dominate males who were higher ranking individually. All accounts of coalitions in male baboons agree that frequent allies tend to be older males who are usually long-term residents (DeVore, 1962; Hall and DeVore, 1965; Ransom, 1971, 1981; Packer, 1979a; Rasmussen, 1980; Collins, 1981).[7] Collins (1981) found that the two highest-ranking males, both of whom were young adult newcomers, tended to gain consort partners through single challenges and to lose

[7] Because age and length of residency tend to be positively correlated, it is difficult to determine the extent to which each factor affects coalition formation independent of the other. Qualitative observations suggest that both factors are important. Collins (1981) and Rasmussen (1980) reported that the most recent transfers in their troops did not form coalitions, while resident males of equivalent age did. Among olive baboons at Gombe, however, some older males formed effective coalitions soon after entering a new troop (Ransom, 1981). In EC in 1983, young adult long-term resident males (i.e., natal males) formed coalitions more often than newcomers of equivalent age but less often than did older long-term resident males (Smuts and Watanabe, in preparation).

females to coalitions, whereas the reverse pattern characterized longer term resident males. Three studies reported that coalitionary challenges were more common and had higher rates of success than single challenges (Ransom, 1971, 1981; Rasmussen, 1980; Collins, 1981). In all of these studies, coalition formation helped to explain the fact that some longer term residents and/or older males had higher consort activity than expected based on their agonistic ranks. For example, Rasmussen (1980) observed one middle-aged male whose agonistic rank was 8 of 13. He ranked 7 in consort success when considering only consort partners that he obtained on an individual basis. However, "the number of females he acquired with the aid of coalitions gave him a total score that was second only to the highest-ranking male" (Rasmussen, 1980, pp. 7:24–7:25).

These results suggest that in many different troops, older residents rely on alliances with one another to take females away from younger, higher-ranking males. However, these findings also raise a problem. Based on observed consort takeovers in EC and other troops, young, high-ranking males very rarely succeed in taking females away from other males. Yet in EC in 1983, the two most successful consorters were young adults. The inconsistency is even greater for the troops studied by Rasmussen (1980) and Collins (1981), because in these troops, young adult, high-ranking males averaged higher consort activity than older, lower-ranking males. How did these young males acquire females? One possibility is that high-ranking males obtained most of their consort partners at night, when observers were not present to record the takeover, and maintained the consort relationship into the following morning, when observers were most likely to collect data. This hypothesis is supported by Rasmussen's (1980) study showing a significant, positive correlation between male rank and the frequency with which males obtained consort partners at night (based on inferred takeovers). It is tested further below.

Comparison of Consort Turnovers during the Day and at Night

As before, when consort turnovers were not observed directly, they were inferred on the basis of observations of two or more males in consort with the same female within a brief timespan. Daytime takeovers were inferred only when successive observations of different consort partners occurred less than 3 hours apart. Turnovers at night were inferred only when consort records were available after 1600 one evening and before 0800 the next morning. This is a conservative method, and the actual number of nightly turnovers was no doubt considerably higher than indicated by the proportion of nightly turnovers in this sample.

Figure 7.7 shows that during the day, most turnovers resulted in a takeover by a male who ranked lower than the previous consorting male, whereas at night, most turnovers resulted in a takeover by a male who ranked higher than the previous consorting male. This difference was highly significant for both study periods [*3*].

In order to examine the relationship between timing of turnovers and ages of rivals, turnovers that involved only males from different age/residence classes were considered (Figure 7.8). Consistent with the previous results on rank, at night older residents lost females to younger males, whereas during the day younger males lost females to older residents; these differences were significant for both studies [*4*].

Avoidance of Competition with Males of Same Age/Residence Class

The data revealed a third important pattern: Males tended to take females away from males of a different age/residence class more often than expected and to take females away from males of the same age/residence class less often than expected (Figure 7.9). These differences were significant for both studies [*5*]. Among young adults, the tendency to avoid direct competition with one another was also apparent in their relationships with females, both in and outside of consortships. Nineteen pairs of males shared a female Friend. Eleven of these pairs involved two older residents, seven involved an older resident and a younger male, and only one involved two younger males. A similar pattern was evident in the pairs of males who "shared" a consort partner (males who were seen to consort with the same female during a conception cycle). Eighty-seven such pairs were recorded. Forty-four (50%) involved older residents and younger males, 38 (44%) involved two older residents, and only 5 (6%) involved two younger males.

Changes in One Male's Tactics over Time

The results suggest that male baboons pursue two very different strategies, depending on rank and age/residence characteristics, which tend to covary. Low-ranking, older residents take females away from higher-ranking, young adult males during the day by forming coalitions with males like themselves, whereas high-ranking, young adult males take females away from lower-ranking, older residents at night on their own. If this hypothesis is correct, then individuals should show a change in tactics over their life history. This was indeed the case for Sherlock, who shifted from a young, recent transfer of high rank (5 of

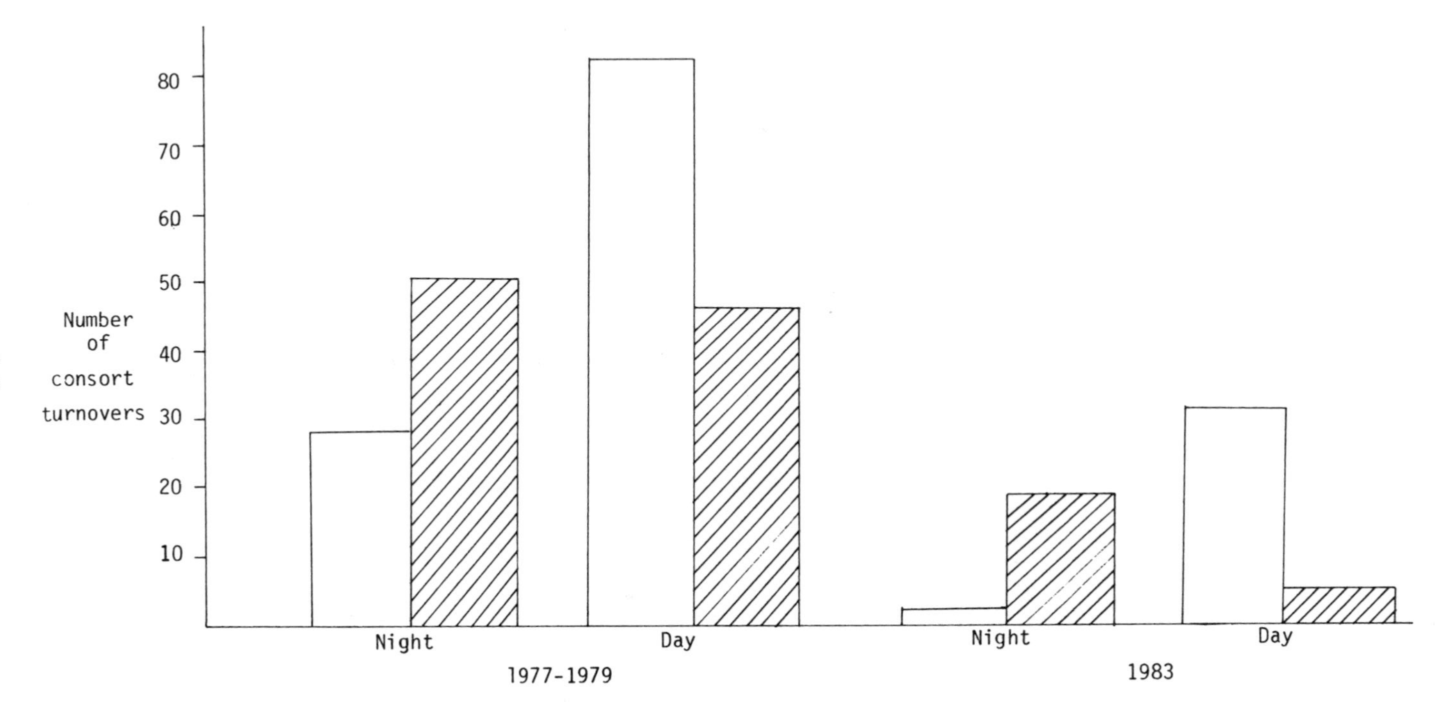

Figure 7.7. Relationship between agonistic rank and consort takeovers at night and during the day. Hatched column, higher-ranking male takes female from low-ranking male; unfilled column, lower-ranking male takes female from high-ranking male.

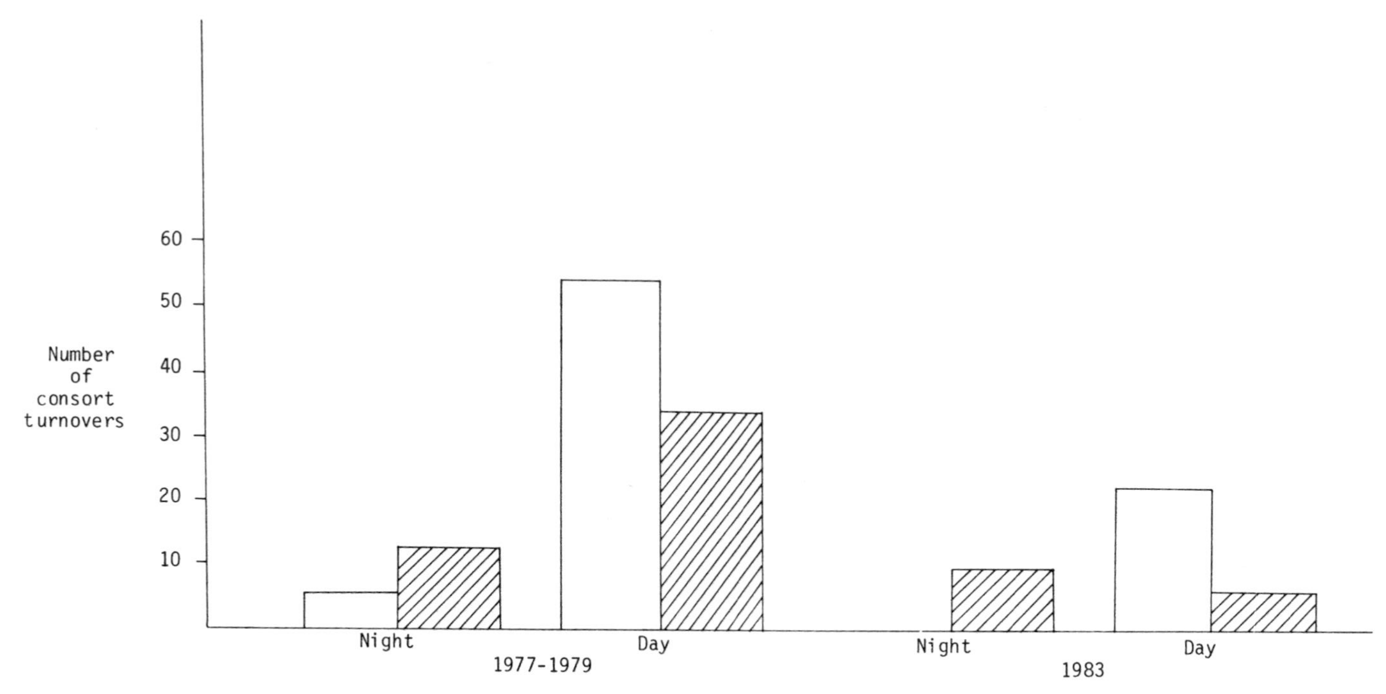

Figure 7.8. Relationship between age/residence status and consort takeovers at night and during the day. Hatched column, young male takes female from resident male; unfilled column, resident male takes female from young male.

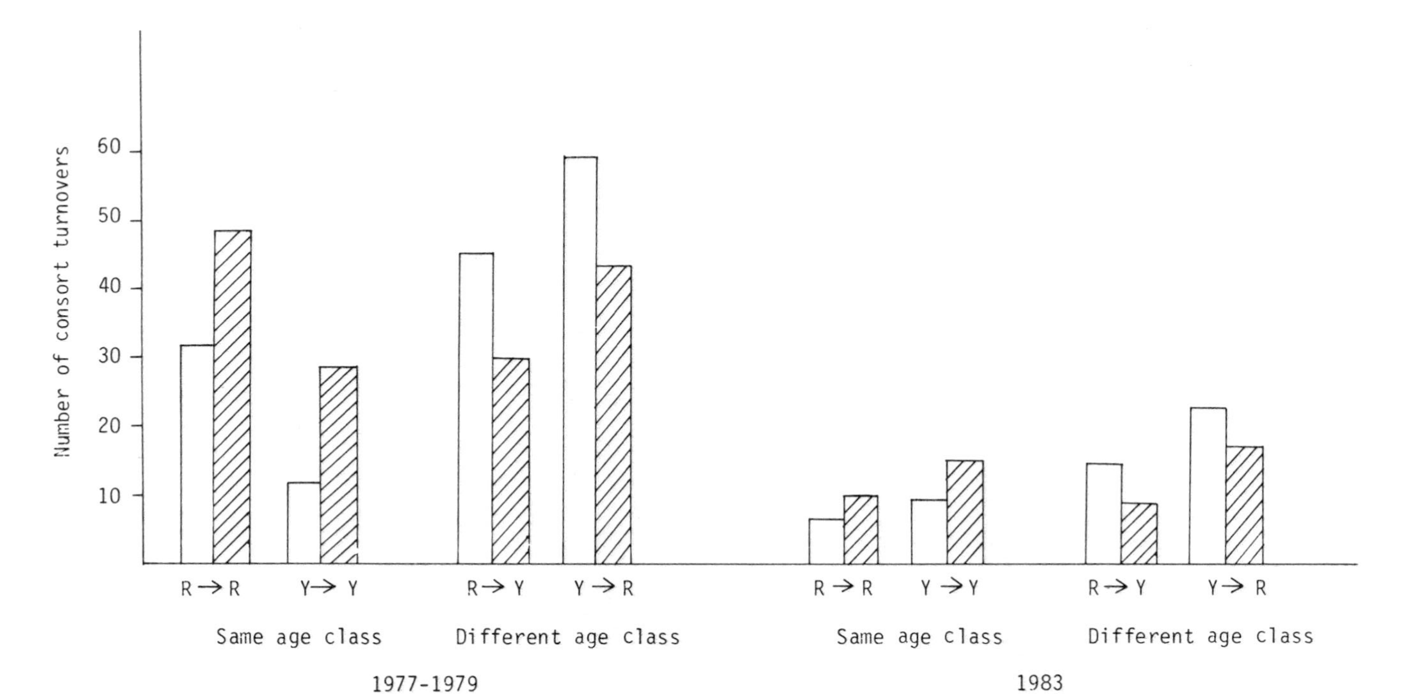

Figure 7.9. Observed and expected frequencies of consort takeovers between males from the same age/residence class and between males from different age/residence classes. R, older, long-term resident male; Y, young adult male. Expected values are based on the total numbers of wins and losses by males of each age/residence class. Hatched column, expected; unfilled column, observed.

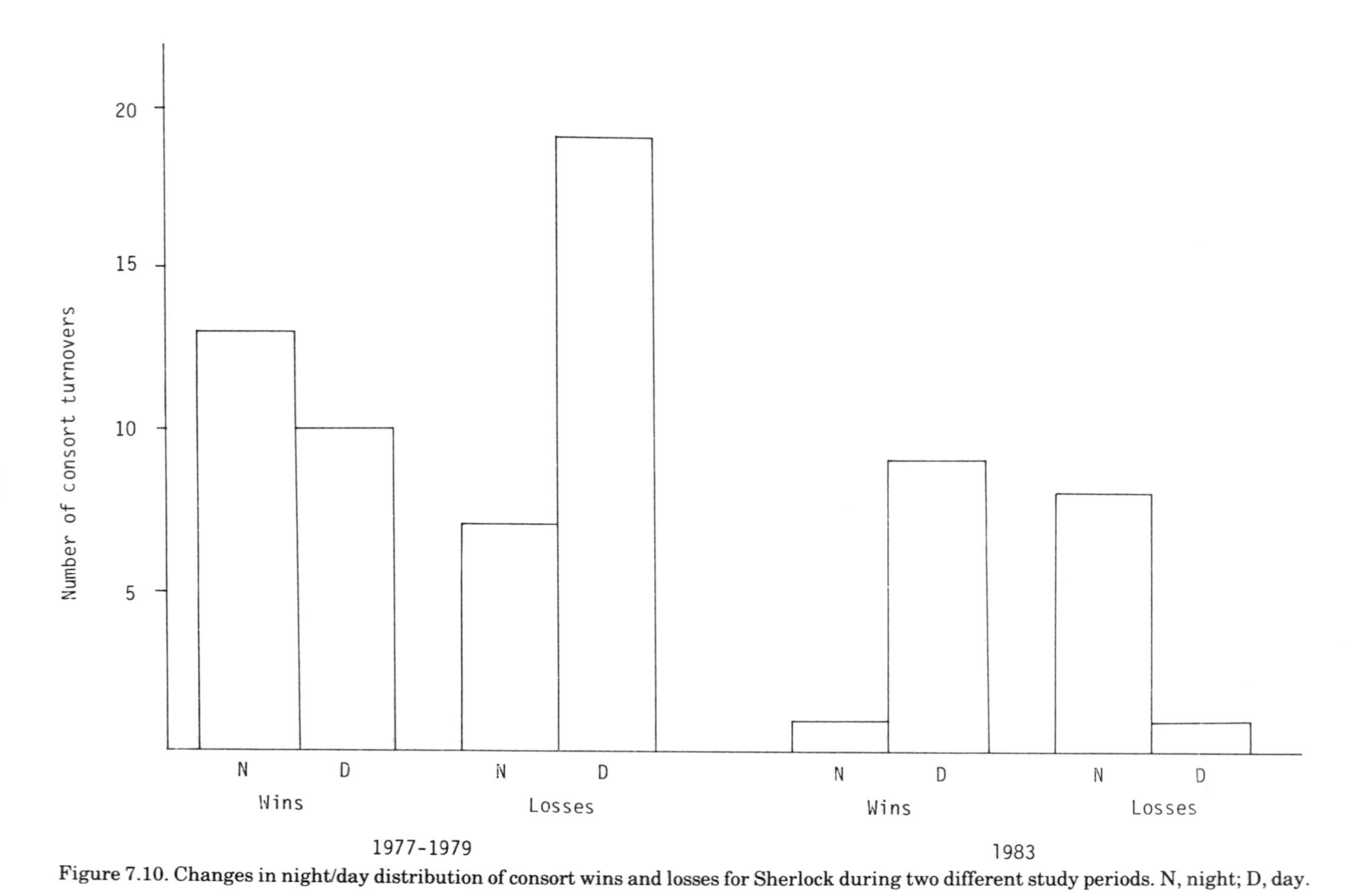

Figure 7.10. Changes in night/day distribution of consort wins and losses for Sherlock during two different study periods. N, night; D, day.

13 males) to an older resident of low rank (12 of 12 males) between the two studies (Figure 7.10). During Study 1, Sherlock won females at night and lost them during the day [*6*], and during Study 2 he won females during the day and lost them at night [7]. In his youth, when Sherlock won a female away from another male, it was nearly always from an older resident (87% of 23 turnovers), and when he lost a female it was to these same males (85% of 26 turnovers). The reverse pattern was evident in Study 2. All 10 of Sherlock's wins involved younger, higher-ranking males, and all but one of his losses involved these same males. Observations of Sherlock taking females away from other males are not available for Study 1, but in Study 2, four out of his six observed takeovers were accomplished with the help of allies. The other two takeovers were opportunistic (see Table 7.4).[8]

Discussion: Age-Dependent Tactics and Coalitions

The results reported above raise a number of questions:

Why do young, high-ranking males tend to acquire females at night, whereas older, low-ranking males obtain them during the day? Two factors seem to give young males attempting single takeovers an advantage at night. First, younger males tend to be more agile in the trees or on sleeping cliffs than older males, allowing them to outmaneuver their competitors (Collins and Nash, personal communication cited in Packer, 1979b; Rasmussen, 1980; Collins, 1981; personal observation). Second, because of reduced visibility at night, it is probably more difficult for males to monitor events associated with consort turnovers, reducing the chances that the solo challenges of young males will result in an opportunistic takeover by a third party.

Older resident males, in contrast, appear to suffer a disadvantage at night. Successful coalitionary challenges rely on tightly orchestrated, synchronized movements among allies. This coordination of action appears to be difficult to achieve at night, both because of poor visibility (Popp, 1978) and because of the complex, three-dimensional sleeping environment (Rasmussen, 1980). Thus, the primary tactic of older males may be of limited value at night.

[8] Two other males, AA and BZ, were also present in EC during both studies. They could not be used to analyze how changes in age/residence status affect competitive tactics because BZ was classed as an older resident during both studies, and AA almost never formed consortships as a young male during Study 1. As an older resident during Study 2, AA's consort turnover profile was very similar to Sherlock's. The same was true for BZ.

Why are there so few successful single challenges by high-ranking males during the day? One reason may be that single challengers are particularly vulnerable to opportunistic takeovers. During Study 2, we observed young adult males following an older resident in consortship on 22 different occasions. During more than one-half of these occasions (59%), a second older resident male persistently shadowed this trio. Since the older resident was never observed to challenge the consorting male himself (see below), it seems likely that he was following the others in order to increase his chances of effecting an opportunistic takeover once the younger male challenged the consortship. During Study 2, one-third of the solo challenges by young adult males resulted in an opportunistic takeover by an older resident who had been monitoring the consorting pair, and two of the three solo challenges in which the young adult showed aggression toward the consorting male were exploited in this way. These results suggest that a significant proportion of single challenges by young males during the day might actually benefit an older resident rather than the challenger—particularly if the challenger uses aggression. Only 1 of the 13 coalitionary challenges resulted in an opportunistic takeover (by another older resident), which suggests that males adopting this tactic are less vulnerable to exploitation by an uninvolved bystander.

Why do males obtain most of their consorts from males of a different age/residency class? The answer to this question seems relatively straightforward for young, high-ranking males. Other young, high-ranking males are formidable opponents capable of inflicting serious injury. Older males are less risky foes, in part because their canines are usually broken, worn, or both (Packer, 1979b). None of the single challenges directed to older residents during Study 2 was observed to result in injuries to the young rival, but during a rare challenge by a young male against another young male, the challenger received a deep canine slash wound to the cheek that bled profusely and took several weeks to heal.[9]

[9] The reluctance of young adults to challenge one another one-on-one may help to explain an otherwise puzzling observation. In EC, young adult males frequently attempted to initiate coalitionary challenges against other young adult, consorting males by threatening the consort while simultaneously soliciting aid from two or more older residents. Although this technique sometimes succeeded in provoking a consort turnover, the young adult did not appear to benefit from his actions, since an older resident always acquired the female. However, once the female was in consort with an older resident, the young male who initiated the takeover was in a better position to accomplish

The tendency for long-term resident males to take females mainly from young, high-ranking males rather than from older, less intimidating opponents, is less easily explained. One possibility is that it is difficult for an older male to form an effective coalition against another older male because this leaves him with one less potential ally. However, Rasmussen (1980) and Collins (1981) found that coalitions of three males were nearly always successful in effecting a consort turnover. This means that the coalitionary tactic continues to be an effective option as long as there are at least three remaining allies. However, during Study 2, when there were four older residents in the troop, none of these males was ever seen to form a coalition against one of the others or to challenge one of the other's consortships singly. Rasmussen (1980) also reported that frequent coalition partners did not form coalitions against one another. What is the reason for this?

Perhaps allies refrain from challenging one another in order to preserve a relationship that is mutually beneficial. Packer (1977) suggested that male baboon coalition partners engage in reciprocal altruism (Trivers, 1971): One male helps another to take a female away from a third male, and, at a later date, the male he helped is likely to help him in turn. This hypothesis is supported by the observation that the same males tend to form repeated coalitions with one another (DeVore, 1962; Hall and DeVore, 1965; Packer, 1977; Collins, 1981; see also Table 7.4). Three other observations support the reciprocal altruism interpretation. First, the more often a male joins coalitions, the more successful he is at enlisting the aid of others (Packer, 1977). Second, each participant in a stable coalition wins the female some of the time (Table 7.4; Packer, 1977). Third, Rasmussen (1980) showed that the probability that a participant in a coalitionary turnover would win the female was a direct function of his overall rate of participation in coalitions. This held true in EC as well: Males who rarely participated in coalitions never won the female, while all four of the regular participants did.

Thus, male baboons derive important benefits from the development of stable, cooperative relationships with other males. Evolutionary theory predicts that such relationships are most likely to survive when each partner follows the "tit-for-tat" strategy, responding to an altruistic act with another altruistic act and responding to defection (an act that inflicts a cost) by defecting also (Axelrod and Hamilton, 1981). Packer's analysis (1977) of reciprocal altruism in male baboons dem-

---

a second takeover on his own, and this may be why young adults sometimes supported the efforts of older males to take females away from their peers.

onstrated repayment "in kind." However, there is no *a priori* reason why reciprocal altruists should exchange only one type of benefit (Seyfarth and Cheney, 1984). If male baboons follow the behavioral rules predicted by theory, then frequent coalition partners will strengthen their relationship—and thus the benefits each acquires—by refraining from acts that inflict a direct and obvious cost on the partner, such as challenging his consortships.

Relationships among male savannah baboon coalition partners resemble relationships among male hamadryas baboons belonging to the same band. In both cases, males are inhibited from contesting "possession" of a female belonging to another male, and they cooperate in acquiring females and protecting them from rivals (Kummer *et al.*, 1974; Abegglen, 1984; Kummer, 1984). Kummer (1982, p. 124) has suggested that emotional bonds between two males may prove to be a "prementai mechanism for reciprocal altruism" among male baboons. In hamadryas baboons, males remain in their natal band, and so male–male bonds are based on long-term familiarity and, in some cases, on kinship (Abegglen, 1984; Kummer, 1984). In savannah baboons, male–male bonds usually involve unrelated individuals who meet for the first time as subadults or adults when they transfer from their natal troops into new ones (Packer, 1977; observations from EC). Establishing a stable, cooperative relationship with an unfamiliar partner probably requires considerable time and energy (Kummer, 1982; Smuts and Watanabe, in preparation). This may be one important reason why, in savannah baboons, older, long-term residents form effective coalitions much more often than do younger males.

Because of the time and energy required to form cooperative relationships, each coalition partner becomes an important investment for the other partner (Kummer, 1982). We would therefore expect males to protect this investment by defending their partners against potential injury and by frequently reaffirming the relationship. Observations of two older EC males, AA and BZ, provide evidence for both behaviors. Their coalitionary relationship began to develop in 1977, making it the oldest supportive relationship among EC males. Twice during Study 2, BZ, who appeared to be a more effective fighter than AA, was observed defending his partner against severe aggression by a young, high-ranking male. During one of these occasions, BZ heard AA scream from over 50 m away, immediately ran to the scene, and hurled himself through the air to land on top of the younger male who was attacking his partner. Their unusually close relationship was also reflected in their greetings (Smuts and Watanabe, in preparation). Each morning, as soon as the baboons began to stir, AA or BZ would seek out the other for a relaxed, leisurely greeting. One of them would present his rear to the other who would then grasp his partner's hips or gently touch his genitals while both males lip-smacked and gazed into one another's faces. Inevitably, a few moments later, they would greet again, reversing the roles of the previous greeting: This time, the male who had received the present would present to his partner. Their reciprocal greetings, in which each male took turns adopting the conciliatory role of presenter, seemed to mirror their coalition relationship, in which each took turns helping his partner to acquire females from young males. Reciprocal greetings and genital touching were rarely observed among other EC males and only in obviously tense interactions (Smuts and Watanabe, in preparation).

Why does agonistic rank predict consort activity in some troops but not in others? In EC, high rank gave males a competitive advantage at night but not during the day, indicating that the benefits of high rank are conditional rather than absolute. The ability of older, low-ranking males to form coalitions decreases the advantages of high rank, and effective coalition formation, in turn, will depend on demographic and environmental parameters. For example, any factor that reduces the number of long-term resident males in a troop, such as small troop size, high male mortality, or high rates of emigration by nonnatal males, is likely to reduce the probability that effective coalitions will form. In one yellow baboon troop at Amboseli (Hausfater, 1975) and two chacma troops in Botswana (C. Busse and S. Smith, personal communication), male–male coalitions were very rare and consort activity was positively correlated with agonistic rank. All three troops were considerably smaller than EC, and there is some evidence that male mortality rates and/or emigration rates were also higher, although systematic comparisons among these troops have not yet been made.

The type of terrain is a second factor that might affect coalition formation. Rasmussen (1980) suggested that in areas where vegetation and/or rocks break up the environment and obscure vision, it may be more difficult for coalition members to perform the synchronized movements necessary to provoke a consort turnover.

A third factor that may affect the correlation between agonistic rank and consort success is the degree to which female choice strengthens this relationship or acts in opposition to it. In EC, female choice generally favored older residents, because females preferred to mate with their male Friends (see Chapter 8), and older residents had more Friends than did higher-ranking, young males (Chapter 4). Rasmussen

(1980) pointed out that in troops undergoing high rates of male turnover, females will have fewer opportunities to form long-term bonds with males, and they may instead prefer high-ranking males as mates. Three studies provide evidence that females sometimes prefer to mate with higher-ranking males, and in these troops there was a positive correlation between male rank and consort activity (Seyfarth, 1978a,b; Rasmussen, 1980; Collins, 1981).

Finally, particular individuals and unique events can influence the relationship between agonistic rank and consort activity. For example, the most successful consorters during the two EC studies had radically different profiles: Cyclops was an old, large, very low-ranking male while Dante was a young, small, very high-ranking male. What both males had in common was an extraordinary ability to remain calm in the face of harassment by other males (see below). Sapolsky (1983) found that a single event—the injury of the alpha male in a fight and his resultant loss of rank—was followed by a dramatic change in the tactics male baboons used to compete for females. The frequency of coalitions increased significantly, and high-ranking males had greater difficulty monopolizing individual females during a single estrous cycle.

All of these factors—troop size and demography, terrain, female choice, attributes of particular individuals, and unique historical events—affect the form and outcome of male–male competition for mates, and they are likely to interact in complex ways. It is therefore no wonder that a single aspect of this complex phenomenon, the relationship between agonistic rank and consort activity, is so variable across studies. Recent research indicates that male baboons learn to adopt competitive tactics suited to their individual capabilities and to the particular situation (Packer, 1977, 1979a,b; Seyfarth, 1978b; Rasmussen, 1980; Collins, 1981; Strum, 1982). The winners in such a sophisticated contest are not necessarily those of superior rank, but those who know best how to play the game.[10]

[10]Some authors have questioned the validity of the concept of *agonistic rank* itself. Rowell (1967), Harding (1980), and Strum (1982) reported frequent reversals in the outcome of dyadic agonistic encounters among male olive baboons. As a result, male dominance hierarchies in these troops lacked internal consistency, and all three authors encouraged future researchers to shift their focus away from dominance and toward other aspects of male behavior, especially affiliative relationships. Data from EC (Smuts and Watanabe, in preparation) suggest that reversals in agonistic encounters may themselves provide important information about male competitive tactics. For example, although young adult males normally dominated older residents,

PSYCHOLOGICAL COMPONENTS OF MALE COMPETITIVE TACTICS

When baboon researchers write about determinants of male consort success, we tend to consider factors like those discussed above—factors that can be quantified and compared across studies. But when we *talk* about determinants of consort success among ourselves, more often than not we end up discussing personality traits and individual strategies. Yet because such factors are difficult to quantify, they have received little systematic attention. In this section, I use observations from EC to indicate how individual characteristics may affect the outcome of male–male competition. In this discussion, I speculate about the motives and emotions of the males, because, in many cases, the language of psychology seems the most appropriate medium for communicating about the complex interactions observed in baboons.

Alex, an older, resident male, is in consort with Andromeda. Dante, a young adult, high-ranking male, has been following and harassing the pair all day. As the troop moves toward the sleeping cliffs, Dante begins to intensify his harassment. He slowly circles Alex and Andromeda, threatening Alex with raised brows and repeated yawns that reveal his canines, which are in perfect condition. Alex and Andromeda both ignore Dante's threats and continue foraging. Briefly, Dante too resumes foraging, and Andromeda moves toward him as she looks for a new feeding site. Dante immediately avoids her, but then resumes his threats, following the pair onto the sleeping cliffs. Alex and Andromeda sit on a ledge, and Alex grooms his partner, turning his back to Dante. Dante exploits his youthful

---

older males were sometimes able to win dyadic encounters, and even fights, when the stakes were particularly high (e.g., when competing over access to meat or when defending oneself against a potentially injurious attack). These results are consistent with theoretical predictions that the outcome of a contest will depend not just on the rivals' relative fighting abilities but also on the costs and benefits to each of winning and losing the encounter (Maynard Smith and Parker, 1976; Popp and DeVore, 1979; Bachmann and Kummer, 1980). Observations of repeated encounters between the same pairs of EC males also suggest that participants' assessments of the costs and benefits of each encounter—and therefore the outcome—depend not only on the context but also on the pair's previous history of interaction. Thus, although agonistic *rank* may sometimes fail to be a useful general predictor of behavior, agonistic *interactions* are an important source of information about how males compete and how male–male relationships change over time. These ideas will be developed further in a future publication (Smuts and Watanabe, in preparation).

agility on the rocks by leaping in a circle around the consorting pair. As he circles them, he grinds his canines together, producing a low, grating sound that serves as a constant reminder of his superior weaponry. Alex grooms Andromeda more intensely, but she becomes restless and he herds her away from Dante. Dante repeats his performance, and, each time, Alex appears more nervous. Finally, as Andromeda begins to ascend the cliffs, Alex turns and yawns in threat at Dante. Dante yawns in return, Alex chases him briefly, and then he returns to Andromeda. Dante runs away a short distance but then resumes shadowing Alex. Alex again turns to face Dante and begins to scream defensively. Dante grinds his teeth and stares at Alex. Alex hesitates, looking back and forth between Andromeda, who is watching both males, and Dante, still grinding his teeth. Finally, Alex charges Dante. Dante leaps away but then turns to face Alex. The two males chase back and forth briefly. As soon as the chasing begins, Andromeda runs away from both males. Dante rushes after her and establishes a new consortship. Alex screams for a full minute as he watches Dante move away with Andromeda (focal consort sample, 4 July 1983).

This example illustrates several important features of consort turnovers in savannah baboons. First, a rival will rarely attempt to take a female by approaching her directly or by interposing himself between the female and her consort. In fact, if the female or the consorting male approaches a rival, he will nearly always avoid them (Collins, 1981), as Dante avoided Andromeda, above. Second, rivals usually do not chase or attack the consorting male unless the latter responds aggressively to the rival's harassment. Third, the rival's inhibition against approaching the female is generally maintained until (*a*) the consorting male abandons the female either "voluntarily" or because he responds aggressively to the rival's threats or (*b*) the female runs away from her consort. Often these events coincide, as in the example above.

Why don't rivals attempt to gain the female directly either by approaching her or by attacking the consorting male? The most likely explanation is that both tactics increase the probability of a fight between the rival and the consorting male. It will normally be in the interests of both males to avoid combat for two reasons. First, during a fight each opponent risks injury from the other's canines. Second, once a fight breaks out, the chances of an opportunistic takeover by a third male sharply increase. Thus, for each male, the goal is to resolve the contest in his favor without resorting to physical violence.[11]

In this contest, the consorting male has the advantage of possession of the female (all else being equal, it is easier to maintain possession of a resource than to take it away from someone else). The rival, however, has the advantage of maneuverability, because he has no resource to protect. Rivals use this advantage to perform a variety of acts apparently designed to unnerve the consorting male. For example, a rival will sometimes rush at the consorting pair each time the male tries to copulate, repeatedly forcing him to interrupt his mounts to chase off the rival (e.g., Hall and DeVore, 1965; Ransom, 1981). Another tactic is simply to shadow the consorting pair for prolonged periods, gradually inching closer and closer to them. At first, the consorting male tries to ignore the presence of rivals, but as time passes and harassment persists, he often becomes increasingly tense and finds it difficult to maintain an unconcerned posture (Ransom, 1971, 1981).[12]

Shadowing and harassment by rivals is most likely to lead to a consort turnover when several rivals ally together to form a united front (Ransom, 1971, 1981). In one troop, 14 of 16 challenges by three or more allies resulted in a consort turnover (Rasmussen, 1980), and in another troop, all nine such challenges were successful (Collins, 1981). The example below illustrates the types of maneuvers allies use to provoke a consort turnover.

Early in the morning, Dante is in consort with Andromeda. Three older males, Alex, Sherlock, and Zim, are following and harassing Dante. Their movements are so perfectly synchronized that they take on an almost dance-like quality. Sherlock and Zim stand side by side facing Dante and, in unison, they rapidly and repeatedly threaten Dante with raised brows and then glance at Alex, 20 m away, soliciting his aid. Alex lopes over to them, places one arm around Sherlock's shoulder, and all three pant-grunt at Dante in an antiphonal chorus. In one smooth motion Zim lip-smacks, touches Alex's rear, looks at him, grunts at Sherlock, and then circles around to the other

[11] Usually, this goal is reached; recall that out of 21 consort takeovers listed in Table 7.4, only one involved a fight between the consorting male and his opponent.

[12] The female, too, is often unnerved by the rival's presence, and she may respond by continuously moving away from both the rival and her partner, or even by running away. These responses, which increase the chances that a consort turnover will occur, are discussed further in Chapter 8.

side of Dante. When he is opposite Alex and Sherlock, he resumes threatening Dante and, at precisely the same instant, they do the same. Alex embraces Sherlock and, together, they circle Dante and join Zim. All three stand in contact and swivel as a unit to face Dante, who avoids them. Dante appears increasingly tense. He repeatedly interposes his body between Andromeda and the other males and then herds her away by shoving her from behind. Each time he pushes her, Andromeda squeaks in protest. She too seems tense, glancing back and forth between Dante and the other three males. A few minutes later, a fifth male, Boz, appears on the hillside above the consort pair. Alex, Sherlock, and Zim immediately solicit Boz's aid against Dante. Boz runs toward them, and at the same time the other three once again move toward Dante. Dante and Andromeda break away from each other and run in opposite directions. Zim and Sherlock chase Dante while Boz and Alex run after Andromeda. Alex reaches her first, and she stops running and lets him copulate with her. A new consortship is formed (focal consort sample, 2 July 1983).

Whether the rival is acting alone or in consort with other males, his behavior seems designed to manipulate the consorting male's emotions until, finally, the mounting tension and frustration explode into physical action and he either charges the rival(s) or runs away. During the resulting chaos, a consort turnover is likely to occur. Although contests for females occasionally erupt into violence, the battle is usually resolved through psychological warfare, not combat: The rival attempts to break the nerve of his opponent without provoking an attack, and the consorting male tries to maintain his equanimity in the face of repeated harassment.

Consort activity will be influenced by any factors that affect the outcome of this psychological contest, including the prior relationship between rivals and characteristics such as intelligence, confidence, and experience. In baboons, as in people, such characteristics vary widely among individuals. For example, some males seem much calmer around other males than do others. These differences can be quantified by considering adult male responses to the proximity of other males in the context of routine foraging. During focal samples of adult males, I recorded an "approach" each time a male moved to within 2 m of another. Sometimes males responded to such approaches by avoiding the other male and sometimes by initiating a tense greeting through stylized gestures (Ransom, 1981; Smuts and Watanabe, in preparation). In other cases, however, the approach produced no reaction or evoked only a glance; both were considered a "nonresponse." For each male, I determined the rate at which he evoked nonresponses in other males and the rate at which other males evoked nonresponses in him; the first measure indicates how relaxed other males were around him, and the second reflects how relaxed he was around other males. Both measures were positively, but not significantly, correlated with male consort scores during conception cycles. Since the two measures were significantly correlated with one another [*8*], they were combined to increase the sample size. This new measure, which reflected each male's total rate of involvement in nonresponse interactions regardless of whether he was the approacher or approachee, was positively and significantly correlated with male consort scores [*9*]. Thus, the more successful consorters tended to be males who were relaxed around other males and males around whom other males were relaxed during the course of routine activities. This suggests that individual characteristics that determine nonresponse rates, such as experience and a calm disposition, also affect the outcome of the intense psychological warfare that characterizes competition for mates.

This hypothesis is supported by the behavior of Cyclops, who was the most successful consorter in EC during Study 1 (see Figure 7.1). Cyclops had the highest nonresponse rate of any male, and he also demonstrated a superior ability to remain calm when threatened by other males. On 11 occasions, I saw an adult male persistently threaten Cyclops at very close range (a total of eight different males performed the threatening role). In each case, the threatening male was of a higher agonistic rank than Cyclops and had superior canines. These threats were intense and prolonged, involving repeated circling, tooth-grinding, and canine yawns that flashed the younger male's canines within inches of Cyclops's face. Cyclops did not react to these threats in any way discernable to the observer, and, eventually, the higher-ranking male abandoned the attempt to provoke a response.

SUMMARY

Some EC males consorted with estrous females much more often than did others. The most successful males were not necessarily high ranking: Older long-term residents, who were usually low ranking, had higher consort scores on average than did young adult males, who were usually high ranking. The male with the highest consort score showed distinct mating preferences, favoring those females who were most likely to conceive and those who were not affiliated with his closest rivals.

Analysis of successful consort takeovers showed that males pursued two very different strategies, depending on their rank and age/residence status. Low-ranking, older residents took females from higher-ranking, young adult males during the day by forming coalitions with males like themselves. High-ranking, young adult males, in contrast, took females from lower-ranking, older residents at night on their own. Young adult males avoided challenging other young adults, probably because of risks of injury. Older residents also avoided challenging one another; this inhibition may reflect allies' attempts to preserve mutually beneficial cooperative relationships.

Whether rivals acted alone or in concert, their behavior seemed designed to provoke the consorting male into an active response that would decrease his proximity to the female and increase opportunities for a consort takeover. The consorting male, in turn, attempted to ignore his rivals, to stay close to his partner, and to retain a calm demeanor. Although such contests occasionally involved combat, they were usually resolved without violence, and psychological skills often appeared to be more important than physical prowess in determining the outcome.

These results help to clarify the nature of mating competition among EC males, but only one-half the story has been told. The other one-half concerns the estrous females over whom males compete so intensely. Female baboons manifest strong mating preferences, and, as we will see in Chapter 8, these preferences influence the form and the outcome of male–male competition in several important ways.

NOTES ON STATISTICS

[*1*] Comparison of cumulative distribution of consort scores for conception versus nonconception cycles. Kolmogorov-Smirnoff two-sample test, $d = .12$, $n_1 = 127$, $n_2 = 163$, $\chi^2 = 4.07$, d.f. = 2, n.s.

[*2*] Correlation between male consort rank and male agonistic rank. Spearman rank correlation coefficient, $r_s = -.42$, $N = 13$, n.s.

[*3*] Comparison of the frequency of consort takeovers by a high-ranking male from a low-ranking one versus those by a low-ranking male from a high-ranking one, during the day and at night.

Study 1: $\chi^2 = 12.0$, d.f. = 1, $N = 148$, $p < .001$

Study 2: Fisher exact probability test, $N = 50$, $p < .001$

[*4*] Comparison of the frequency of consort takeovers by older residents from young adults versus those by young adults from older residents, during the day and at night.

Study 1: $\chi^2 = 3.85$, d.f. $= 1$, $N = 104$, $p < .05$

Study 2: Fisher exact probability test, $N = 35$, $p = .0001$

[*5*] Comparison of the frequency of consort takeovers between males from the same age class versus those between males from a different age class (takeovers during the day and at night combined).

Study 1: $\chi^2 = 25.2$, $N = 148$, $p < .001$

Study 2: $\chi^2 = 5.9$, $N = 50$, $p < .02$

[*6*] Comparison of the frequency of consort takeovers involving Sherlock in which he won the female versus those in which he lost the female, during the day and at night (Study 1). $\chi^2 = 3.29$, d.f. $= 1$, $N = 49$, $p < .025$ (one-tailed)

[*7*] Comparison of the frequency of consort turnovers involving Sherlock in which he won the female versus those in which he lost the female, during the day and at night (Study 2). Fisher exact probability test, $N = 19$, $p < .005$

[*8*] Correlation between the rate of nonresponse interactions in which the male was the approacher and those in which he was the approachee. Spearman rank correlation coefficient $r_s = .71$, $N = 11$, $p < .05$

[*9*] Correlation between the rate of participation in nonresponse interactions and consort rank. Spearman rank correlation coefficient $r_s = .50$, $N = 11$, $p < .05$ (one-tailed)

8 BENEFITS OF FRIENDSHIP TO THE MALE

Adult male Handel copulates with his friend, Dido, whose lips are pursed as she begins to utter a series of loud grunts—the "copulation call." Handel simultaneously threatens a nearby opponent by staring and flashing the white skin above his eyes. Handel had an advantage when competing for mating opportunities with Dido, because, like most females, she favored her Friend as a mate.

INTRODUCTION

Nearly all of the adult and subadult males in EC had friendships with one or more adult females. The only exceptions were the newcomers, who showed strong tendencies to develop friendships with females but had not yet had sufficient time to do so (see Chapter 9). Although females usually took more responsibility for maintaining proximity between Friends than did males, males clearly contributed to these spatial associations (see Chapter 5), and they often took a more active role than females in initiating new friendships (Chapter 9). We have seen that males defended their Friends and their Friends' offspring against aggression by other baboons and sometimes against predators as well (Packer, 1980; Ransom, 1981). This chapter examines why males expend considerable time and energy to form and to maintain such relationships with females, often exposing themselves to risky situations in order to protect their vulnerable associates. What benefits do males derive from these relationships? As in Chapter 6, which considered the benefits of friendships to females, I use the term *benefit* here in the evolutionary sense to refer to reproductive advantages. There are three possible sources of such benefits:

1. *A close bond with a female might allow the male to increase the fitness of offspring he has already sired.* If the infant of a male's female Friend is his infant as well, then male behaviors that increase the mother's ability to care for the infant (e.g., protecting her from harassment by other baboons) or that directly increase the infant's chances of survival (e.g., protecting it from predators) will increase the male's own reproductive success. Several authors have considered the hypothesis that certain types of male–infant interactions in savannah baboons represent such paternal investment (Packer, 1980; Stein, 1981; Busse and Hamilton, 1981).

2. *A relationship with a female might increase a male's chances of siring offspring in the future.* If females and their immature offspring benefit from male investment, and if a female tends to mate around the time of ovulation with males who have previously demonstrated their willingness and ability to invest in her and her offspring, then a male who forms friendships with females may increase his reproductive success in the future even when the female's current infant is not his own. This possible benefit of friendship, first proposed by Seyfarth (1978b), has not yet been examined in detail.

3. *A relationship with a female might increase a male's opportunities to "use" her infant during agonistic interactions with other males.*

Several researchers have reported that by carrying an infant, a male can buffer himself against aggression from another male (Popp, 1978; Packer, 1980; Stein, 1981; Collins, 1981; Strum, 1984). Packer and Stein have argued that males form relationships with infants in order to use them in this manner. If this hypothesis is correct, then increased opportunities to develop close bonds with infants could represent an important benefit to males of friendships with mothers.

In the first part of this chapter, I present data from EC relevant to the first two possible benefits. Four questions are addressed. First, do males tend to have friendships with pregnant and lactating females with whom they consorted in the past, that is, with females who are the mothers of infants they might have sired? Second, when a female resumes sexual cycles, is she more likely to consort with males who were her Friends during the preceding period of pregnancy and lactation? Third, how do female baboons express and exert their mating preferences? Fourth, how important are paternity and friendship with the infant's mother in determining male–infant relationships?

In the second part of the chapter, male carrying of young infants during agonistic encounters with other males is considered. Several possible explanations for this behavior are evaluated. This discussion returns to the issues stressed in Chapter 6: the ever-present danger of male aggression and the ways in which males both exploit and protect their more vulnerable associates.

FRIENDSHIPS AND PREVIOUS CONSORT ACTIVITY

Probable Fathers and Friendships

Do males associate with the mothers of infants they might have sired? To answer this question, the consort records described in Chapter 7 were used to determine probable fathers of infants conceived during the first 7 months of the study—that is, infants conceived early enough so that data were available for the mothers' sexual cycles, subsequent pregnancies, and periods of lactation. For these 15 females, the male with the highest consort score over days D-5–D-1 (a period that reasonably brackets the time of conception) was considered the "likely" father. The male with the second highest score for this period was considered a "possible father." Together these males constitute the "probable fathers." For most females, there were gaps in daytime consort records, and for all, there were gaps during the night. Probable fathers, therefore, represent educated guesses based on observed consort behavior; in all cases, another male could be the actual father.

However, the available consort records probably provide a reasonable estimate of males who consorted with the female at the time when conception was most likely to occur.

This conclusion is supported by paternity exclusion tests on some infants, based on analysis of blood samples (D.G. Smith, unpublished data). In this analysis, protein polymorphisms, which reflect underlying genetic differences, were examined to exclude some males as possible fathers of infants. Unfortunately, analysis revealed insufficient genetic diversity to identify the actual fathers of most infants. In addition, blood samples were unavailable for 6 of the 14 adult males, including 2 males who were frequent consorters (AG and BZ), one who occasionally consorted (AO), and three who were rarely observed in consort with females (IA, JS, and TN). However, for 8 of the 15 infants conceived after the study began, it was possible to exclude one or more fathers, for a total of 26 exclusions. No "likely fathers" were excluded, although in two cases, a "possible father" was excluded. The 4 males (AA, AC, HC, HM) who ranked lowest in consort activity (out of the 8 males for whom blood was available) were excluded a total of 19 times, whereas the 4 males (CY, HD, SK, VR) who ranked highest in consort activity (out of these 8 males) were excluded a total of only 6 times. For one infant (the daughter of PY), it was possible to exclude as a father all but one of the males who contributed blood; since none of the males without blood samples was ever seen in consort with PY, it is also unlikely that any of them was the father. The one male not excluded was HD, the most likely father based on consort records.

Table 8.1 lists the 15 females, the likely and possible fathers of their infants, and their Friends during subsequent pregnancy and lactation. Eight of the likely fathers (53%) and nine of the possible fathers (60%) were also Friends (57% when likely fathers and possible fathers are combined). For 14 of the 15 females, either the likely father or the possible father or both was also a Friend.

Two males, CY and AG, accounted for most of the exceptions to the general trend that females' Friends were also likely fathers or possible fathers. CY and AG had the highest consort scores across all conception cycles occurring before AG disappeared (see Chapter 7). These results suggest that males who were likely or possible Fathers had friendships with the infant's mother during subsequent pregnancy/lactation *unless* the male had unusually high consort activity, in which case he might or might not associate with her. Three possible explanations might account for this weaker relationship between consort partners and subsequent associations when the consorting male was CY or AG:

Table 8.1. Comparison of Probable Fathers of Infants and Mothers' Friendships with Males during Subsequent Pregnancy and Lactation

Female	Date of conception	Likely father[a]	Possible father[a]	Friends
RH	9/77	HC	BZ, CY	BZ —
MM	9/77	CY	AC	AC AO[b]
PY	10/77	HD	CY	HD —
DL	10/77	HM	—	HM PX
PO	11/77	CY	AG	CY SK
DP	11/77	CY	HD	HD PX[b]
LE	11/77	AG	AC	AG AC PL AS BZ
PH	11/77	AO	BZ	AO BZ
HH	12/77	BZ	SK	BZ SK
CG	12/77	AG	SK, VR	AC SK AS IA
ZD	1/78	CY	HC	HC AO
CB	2/78	CY	AG	HD SK
DD	2/78	CY	AG	CY HD
AM	3/78	BZ	—	BZ AO
OL	3/78	AG	AC	AC AS

[a]Likely and possible fathers based on consortships during conception cycles; see text for details.
[b]Male was thought to be the female's son.

1. Because CY and AG were the most active consorters in the troop, a greater proportion of their time might have been spent either in consortships or actively competing for estrous females. Since pregnant/lactating females rarely associated with Friends when the males were in a consortship or harassing other consortships, AG and CY might have been less available as associates and therefore might have been less effective than other males in providing the benefits of friendship. Some females might have avoided forming friendships with AG and CY for these reasons.

2. It is also possible that CY and AG were unwilling to form friendships with more than a few females because the reproductive payoffs would be greater if the time and energy required to develop and maintain additional friendships were used instead to compete directly for estrous females. This suggestion implies that competing for many estrous females and forming long-term bonds with a few females can be viewed as alternative reproductive strategies. It is important, however, to stress that these two strategies were complementary rather than mutually exclusive, since: (a) all EC males pursued both strategies and (b) friendships contributed to males' abilities to compete for mates, as shown below.

3. Because CY and AG were the probable fathers of several infants, females might have faced more competition from other females in forming relationships with them than they faced in forming relationships with less successful males. If competition with other females in part determined whether or not a female formed a friendship with CY or AG, one would expect higher-ranking females to be the ones who formed friendships with them. Table 8.2 tentatively supports this explanation: All females who had CY or AG as both a probable father of their infant and as a Friend ranked in the top one-third of the female dominance hierarchy. Alternatively, if, as suggested above, CY and AG were willing to form friendships with only some of the females for whose infants they were probable fathers, and if they preferred to form friendships with higher-ranking females, then male choice, rather than (or in addition to) female competition, could explain the results in Table 8.2.

Five of the six females who had CY or AG as the likely father, but who did not have a friendship with the likely father, did form friendships with the possible father (Table 8.1). *Only one of the 15 females (CB) showed no overlap between observed consort partners and Friends.*

Altmann (1980) and Stein (1981, 1984) reported similar results for yellow baboons in Amboseli. In Stein's study, all males who accounted for at least 20% of an infant's time in proximity (within 5 ft) to adult males were considered "preferred" males. Males that were seen copulating with the mother during days D-4–D-1 of the conception cycle were the males most likely to be preferred males (Stein, 1984).

These results suggest two reasons for females to consort with more than one male during peak estrus. First, the existence of multiple consort partners may provide a sort of "insurance," so that if one partner is, for whatever reason, unlikely to become a Friend, there may be another male willing to form a friendship. Second, mating with more than one male may increase the chances that a female will have more than one Friend. The fact that in both EC and in Amboseli most females had two Friends (Chapter 4) suggests that, in general, females do prefer to have more than one Friend.

Although likely and possible fathers tended to associate with females whose infants they might have sired, many male Friends were not probable fathers. When the two Friends thought to be the sons of their female associates are eliminated, 31 female–Friend dyads remain. For 17 of the dyads, the males were likely or possible fathers. For 14 of the dyads, the Friend was *not* a probable father. Of these 14 dyads,

Table 8.2. Dominance Ranks of Mothers for Whom CY and AG Were Likely or Possible Fathers of Infants[a]

Female	Dominance rank	Female was Friends with CY
CY was likely or possible father		
DD	1	+
ZD	2	–
PO	8	+
RH	19	–
PY	23	–
MM	24	–
DP	27	–

Female	Dominance rank	Female was Friends with AG
AG was likely or possible father		
DD	1	–
PO	8	–
LE	10	+
CB	20	–
OL	30	–

[a]Note that the two females who ranked higher than LE, but who were not Friends with AG, were Friends with CY, the likely father in both cases.

5 involved subadult, natal males who rarely consorted with any females, and 8 involved young adult, recent transfers. Only 1 dyad (LE and BZ) involved an older resident male (LE, with 5 male Friends, was an unusually gregarious female). This apparent relationship between male residency, consort activity, and friendships in EC can be summarized as follows:

1. Subadult males living in their natal troop had friendships with unrelated females even when they were not likely fathers of those females' infants. These males had low consort success.
2. Young adult males still residing in their natal troop had friendships with females with whom they had consorted (HM and AO) and also with other females (AO). They had moderate consort success.
3. In the first few months after a male transferred into the new troop (usually as a large subadult or young adult), he did not have

friendships with adult females, and he had low consort success (AA, TN, JS). It is important to note, however, that recent immigrants attempted to form friendships with females (see Chapter 9 for details).

4. After living in the troop for 6 months to 1 year, transfer males began to form friendships with females, including both females with whom they had and had not consorted previously. The males with several friendships (SK with 8, AC with 5, HD with 5) had moderate to high consort success; males with fewer friendships (IA with 1) had low consort success.
5. Older males who had lived in the troop for at least 1 year, but probably much longer, had friendships only with females who were previous consort partners during conception cycles, although they did not necessarily have friendships with all such females. These males had moderate to high consort success (AG, BZ, CY, HC, VR).

The most striking pattern to emerge from this summary is the difference between older, long-term residents and all other males. With one exception (BZ and LE), older residents formed friendships only with females with whom they had consorted during peak estrus of the conception cycle. All young males and all recent transfers, however, were willing to form friendships with females whose infants they were very unlikely to have sired. These results suggest (1) that the opportunity to invest in probable offspring is not the only benefit males gain from friendship and (2) that the benefits to males of friendships with females may vary, depending on male age and residence status, a hypothesis considered further in the conclusion to this chapter.

Overall Consort Activity and Friendships

Although the results described above suggest a strong relationship between previous consort activity and subsequent friendships, especially for long-term resident males, the sample of mother/probable father dyads was too small for statistical analysis. In order to determine whether the association between previous consort activity and friendship was a significant one, I expanded the data set by including information on consort activity during nonconception cycles as well as conception cycles.

Data on consort activity for at least two estrous cycles were available for 25 EC females over a period of 2 years, from September 1977 to September 1979. Each male scored one point each time he was seen in consort with a female on days D-7–D-1 of her estrous cycle; a male could score only one point per day per female. Male consort scores were summed across all females (total points), and each male's percentage of total points was calculated. This value is referred to as his consort probability score. Then, for each female, I calculated the total number of points given to all males over all of her cycles, and the proportion of that total accounted for by each male in the troop. This value is referred to as the male's individual female consort score. Males whose consort probability scores were less than 2% were eliminated, leaving 10 males.

Eight of the 25 females included in this analysis cycled early enough in the study so that data on their relationships with males during subsequent pregnancy and lactation were available. These 8 females and the 10 consorting males produced 80 male–female dyads. Each dyad in which the male's female consort score exceeded his consort probability score was assigned a plus; all other dyads were assigned a minus. A plus, then, indicated that the male consorted with the female on a greater proportion of her cycle days than expected, based on the null hypothesis that a male's probability of consorting with a female was independent of her identity. Of the Friend dyads, 60% had pluses compared with only 28% of the Non-Friend dyads [*1*], indicating that males who had friendships with females after conception were the previous consort partners of these females significantly more often than were Non-Friend males.

FRIENDSHIPS AND SUBSEQUENT CONSORT ACTIVITY

We have seen that males tended to have friendships with those females with whom they consorted in the past, but did a friendship with a female increase a male's chances of consorting with her in the future? This question is addressed using the same two methods employed above; first, friendships are compared with paternity estimates and, second, with overall consorting activity.

Probable Fathers and Friendships

Information on consort activity during conception cycles and on friendships *prior* to the resumption of cycling was available for eight females who resumed cycles toward the end of the study (Table 8.3). For the first four females listed, the likely father of the female's next infant was a Friend during her previous period of pregnancy and lactation, and in one of these four, the possible father was also a previous Friend. However, for the last four females, the likely and possible fathers were males who were not Friends of the female before the onset of cycling.

Table 8.3. Comparison of Probable Fathers of Infants and Mothers' Friendships with Males during Previous Pregnancy and Lactation

Female	Date of conception	Likely father[a]	Possible father[a]	Friends
AM	3/78	BZ	—	BZ AO
LI	7/78	HD	CY	HD VR HM[b]
HH	7/78	SK	BZ	SK BZ
AM[c]	11/78	AO	—	BZ AO
CC	11/78	SK	BZ	CY
AT	11/78	CY	BZ	HC AG[d] HS[b]
JU	11/78	HD	VR	SK
CI	1/79	HD	TN	HC VR

[a]Likely and possible fathers based on consortships during conception cycles; see text for details.

[b]Male was thought to be the female's son.

[c]AM appears twice because she miscarried toward the end of her first pregnancy and conceived again soon afterward. Because she was an adolescent female, she was not used as a focal subject, and her male Friends were identified on the basis of her ad lib grooming records. BZ accounted for 52% of her grooming with males and AO for 35% (N=23 grooming episodes with males when she was not in consort).

[d]AG had disappeared by the time AT resumed cycling.

Of the likely fathers in this sample, 88% involved young adult males compared with only 30% of the likely fathers of infants conceived before March 1978, early in the study (Table 8.1). These differences are significant [2], indicating that the mating success of young adult males increased over the course of the study period, whereas that of older resident males declined. The increased mating success of young adult males was primarily due to two immigrants, HD and SK, who achieved full social integration over the course of the study (see Chapter 9). For two females, CI and CC, the inconsistency between previous Friends and subsequent likely fathers apparently reflected these changes. Both females were Friends with older, resident males but consorted most often with SK or HD. The other two females, JU and AT, however, did not fit this pattern. During pregnancy and lactation, JU was Friends with SK,[1] but when she resumed cycles, she consorted mainly with HD. During lactation, AT was Friends with two older residents, but she subsequently consorted with two different older residents.

[1] JU's infant was conceived before the study began, but biochemical analysis excluded as possible fathers of this infant all males for whom blood samples existed except SK (D.G. Smith, unpublished data).

Three other baboon studies have made similar comparisons between previous affiliative relationships and subsequent consort partners. In those troops, as in EC, some consort relationships were clearly based on prior bonds, but others were not (Seyfarth, 1978a,b; Collins, 1981; Manzolillo, 1982). In the neighboring PHG troop, many females did not consort primarily with previous affiliates, and Manzolillo (1982) argued that these results indicated that a male's chances of mating with a female were not improved by prior bonds. However, the hypothesis under consideration here does *not* predict that females will consort primarily with previous Friends, or that the consort scores of Friends will be higher than those of other males. Male–male competition is known to affect male consort activity (Chapter 7), and, in some female cycles, the effects of male–male competition may outweigh the effects of female preferences. In addition, females may not always prefer to mate with their Friends (see Chapter 9). The hypothesis proposed here (see also Seyfarth, 1978b) is simply that friendship will, on average, increase a male's chances of consorting with a female *above what they would be otherwise*. Comparisons of absolute consort scores of Friends and Non-Friends, while informative, do not provide an adequate test of this hypothesis; what is required is a comparison of male consort activity with female Friends and nonaffiliated females when the effects of other factors are held constant. The results of such an analysis are described below.

Overall Consort Activity and Friendships

Consort records were available for 14 females for whom I also had information on prior friendships. One male, AG, had disappeared before these females resumed cycles, leaving 9 males who consorted regularly with females. The 9 males and 14 females produced 126 dyads. Dyads in which the male consorted with the female more often than expected were assigned a plus, as in the analysis of consort activity and subsequent friendships discussed earlier in this chapter. Recall that expected scores were based on each male's consort activity with all females throughout the study.

Dyads involving previous Friends had significantly more pluses (62%) than dyads involving Non-Friend males (29%) [*3*]. These results allow rejection of the null hypothesis that male consort activity was independent of prior friendships. *On average, friendship in the past doubled the probability that a male would form a consortship with that female in the future.* It is also worth noting that the proportion of pluses among Friends obtained in this "prospective" analysis was very similar to that obtained in the "retrospective" analysis described earlier (60%), indicating that friendship (1) predicted consort activity in the future and (2) reflected consort activity in the past, with about equal strength.

FEMALE CHOICE

Why is male consort activity enhanced by previous bonds with females? The simplest explanation is that females sometimes prefer to mate with their male Friends, and that female preferences affect consort relationships. However, given the vulnerability of female baboons to aggression and injury by males (Chapter 6) and the intensity of male–male competition for mates (Chapter 7), it is reasonable to ask: To what extent are female baboons able to exert their mating preferences?

Compared to some other primates, female baboons in peak estrus have few opportunities to assert their mating preferences directly. They can rarely change consort partners simply by leaving one male and approaching another, as female Barbary macaques often do (Taub, 1980), nor do they use aggression to discourage undesired suitors, as vervet monkey and macaque females sometimes do (Andelman, 1985; Enomoto, 1974). The size difference between adult males and females is much greater in baboons than in macaques or vervets,[2] and the extreme vulnerability of female baboons to male aggression probably helps to explain their restricted expression of mating preferences compared to females in some other species (Ransom, 1971, 1981; Packer, 1979a). The form and intensity of male–male competition in baboons may also constrain female mate choice. Unlike baboons, macaques and vervets are seasonal breeders, which means that many females are sexually receptive at the same time. As a result, more males are able to mate. Perhaps for this reason, fights over females are usually less common, and the elaborate, multimale interactions that typify consort turnovers in some baboon populations have not been observed in these species (e.g., Fedigan, 1982; Berenstain and Wade, 1983; Chapais, 1983c; Andelman, 1985).

Despite these constraints, female baboons still appear to have many opportunities to express and exert their mating preferences (Ransom, 1971, 1981; Rasmussen, 1980; Collins, 1981; Strum, 1982). Because I did not sample females in consort, I do not have systematic data with which to test this hypothesis. Instead, I will illustrate opportunities for

[2] Adult female weight as a percentage of adult male weight: olive baboons: 53% (based on EC weights, see Chapter 2); rhesus macaques: 83% (Rawlins *et al.*, 1984; see Table 10.1); vervet monkeys: 80% (Fedigan, 1982).

female choice through examples.[3] What follows is therefore considerably more speculative than previous analyses, and the ideas presented below should be regarded as hypotheses that require further evaluation.

The three examples that follow show how female preferences and aversions can be expressed during the consortship itself by cooperating, or failing to cooperate, with the partner.

Behavior during Consortships

At noon, Delphi, a young adult female, is in consort with Zim, an older, resident male. During an aggressive encounter, Zim loses Delphi to Vulcan, a young natal male about the same age as Delphi. Zim, Alex, and Boz, three older residents, immediately begin to follow the consort pair. Delphi looks back at them, and Vulcan nervously herds her away. He tries to groom her, but she pulls away and begins to feed. At 1256 Vulcan approaches Delphi and begins to mount her. She jumps away, and he watches her as she resumes feeding. At 1258 he tries to mount her again, placing his hands on her back. Delphi walks away and Vulcan follows, still holding on to her. He maintains this "wheelbarrow" position for several steps, but then Delphi swerves sharply to one side and he falls off. He approaches her again 1 minute later, but she moves behind a large bush before he reaches her. He follows, but Delphi continues to circle the bush, darting quick glances behind her at Vulcan. They both stop moving, on opposite sides of the bush. Vulcan begins to circle in the other direction, and Delphi immediately resumes travelling in the opposite direction in order to avoid him. They circle the bush in alternate directions for several minutes, until finally Vulcan catches Delphi. He tries to mount, but Delphi pulls away. Vulcan gives up for the moment and they feed. The other males are following closely, and at one point Delphi and Boz make eye contact while Vulcan's back is turned. They exchange "come-hither" faces, but without the usual accompanying grunts, and Vulcan remains unaware of the interaction. At 1330, Vulcan again tries to mount Delphi, and again she pulls away. Between 1330 and 1442 Vulcan attempts to copulate 27 times, but each time Delphi refuses. At 1442 Vulcan grooms

[3] These examples are drawn from ad lib observations of consortships during the main study and from both ad lib and focal samples of consortships conducted by J. Watanabe and myself during the summer of 1983.

Delphi for the first time. He grooms her for 1/2 hr and then tries to mount. She refuses the mount, but she does groom him briefly. At 1520 Vulcan tries to mount again, and again she refuses. It begins to rain, the baboons rush to the sleeping cliffs, and we lose sight of the pair. During the 3 hours we followed them, Delphi refused 42 copulation attempts (focal consort sample, 11 July 1983).

Dido, the highest ranking female, is in consort with a young natal male, Adonis; they are not Friends. Dido appears tense, and twice when Adonis takes a step toward her while foraging, she starts nervously. A few minutes later after the troop has moved into an area with many bushes, I see Adonis dashing about frantically, apparently looking for Dido. He climbs a tree and looks all around. He spots Dido crouched behind a bush and runs toward her. With a squeal, she runs away and they disappear into some bushes. I try to follow them, and a few minutes later I come upon Adonis, wandering alone, looking for Dido again. I move off to look for another female, and nearly fall into a large hole abandoned by a family of warthogs and now obscured by tall grass. As I move aside, I sense that there is something alive in the tunnel and I step back. A second later Dido's face appears at the entry to the hole. She is hiding from Adonis. I stay nearby to see what she does. Dido remains in the hole for 10 minutes, periodically peering around and then ducking her head whenever she hears a baboon approaching. Finally, most of the troop moves past her, and she climbs out and follows them (ad lib observation, 4 January 1978).

At 1300 Artemis, an older female, is in consort with Hector, an older resident who is one of her Friends. Hector gets into a fight with another male, and Virgil, another older resident who is not one of Artemis's Friends, rushes in and claims her. The baboons are close together, feeding on grass in an open field. Artemis, with Virgil close behind, stands and surveys the troop. Then, with a decisive movement, she leaves Virgil and begins to trot. She moves rapidly through the troop on a wide, circular trajectory that carries her past most of the adult males. As she nears each one, she stops a few meters away and stands with her sexual swelling facing the male. Each time, Virgil moves between Artemis and the male, and then he approaches the male and presents his own perineum—a gesture males often use when trying to appease a rival (Ransom, 1981; Smuts and Watanabe, in preparation). While Virgil and the other male interact, Artemis darts off to approach still another male, and the preceding events are repeated. For 10 minutes she moves rapidly through the troop, and Virgil, beginning to look tired, tries to keep up with her. Boz and Hector begin to follow them and, by the next day, Artemis is in consort with Boz (the turnover was not observed) (ad lib observation, 10 September 1978).

These examples show that, for a male baboon, acquiring a consort partner is only the first step toward mating success. Whether or not the male derives a reproductive benefit from acquiring a partner depends on the female's behavior (Strum, 1982; see also Takahata, 1982a for macaques). The behaviors illustrated above are not uncommon. Females frequently refuse copulations, and since copulations cannot occur without female cooperation, the female has final control over which males she mates with. Although I came across females hiding in holes or under bushes on only a few occasions, I saw males engaged in prolonged searches for their consort partners many times, suggesting that females frequently tried to escape from their consort partners. Approaching and even presenting to other males is also regularly observed among females in consort, and this behavior may affect consort turnovers. Rasmussen (1980) reported that 27% of such approaches were followed later the same day by a consort challenge by the approached male, and Collins (1981) found that females approached most often the males they frequently associated with outside of consortship.

These types of behaviors can affect consort relationships in two important ways. First, by increasing the costs and lowering the benefits to the consorting male, failure to cooperate may reduce the male's motivation to remain with a female, making a consort turnover more likely (Strum, 1982; Chapais, 1983c). Second, by refusing to cooperate with her partner, the female may signal to other males her lack of preference for him. This information, in turn, can have an important effect on another male's willingness to challenge the consort relationship. In experiments with captive hamadryas baboons, Bachmann and Kummer (1980) found that males were significantly more likely to contest possession of a female when the female exhibited a low preference for her "owner." Thus, even if a female's attempts to approach other males or to escape from her partner do not result directly in a turnover, these highly visible behaviors may affect the likelihood of a future turnover by informing other males of her preferences and aversions.

Seyfarth (1978a) provides a striking example of how a female's preference for a long-term associate can affect male willingness to contest "possession" of that female. In the troop of chacma baboons he studied, there were only two adult males and eight adult females. One male, Rocky, was clearly dominant over the other male, Pierre. When cycling, most of the females showed preferences for Rocky, and he was able virtually to monopolize sexual access to these females. However, one female, Wellesley, showed a strong preference for Pierre during all of her reproductive states. During her conception cycle, she formed a consortship with Pierre, and Rocky was never seen to challenge this consortship, even though Wellesley was the only estrous female in the troop at that time (Seyfarth, 1978a, p. 209).

Data from EC discussed in Chapter 7 provide similar examples of how prior bonds between a male and female can affect male–male competition. Recall that CY, the most successful consorter in the troop, showed no interest in the seven estrous females who were Friends with his three closest rivals. Packer (1979b) found that males tended to favor consort partners who were not preferred by other males, and Collins (1981) noted that even high-ranking males were sometimes reluctant to contest access to females who had frequent affiliative interactions with their current consort partner outside the consort relationship.

Rasmussen (1980, 1983b) has provided a quantitative analysis of the relationship between female behavior during consortship and female preferences for males, measured by the frequency with which cycling females approached males outside of consortship. Before discussing her results, it is important to note that the affiliative preferences revealed by this measure do not necessarily reflect long-term friendships for three reasons. First, some females show preferences for Non-Friends while they are cycling but then revert to their previous preferences for Friends after conception (Seyfarth 1978a,b; see also Chapter 9). Second, when females resume cycling, previous long-term relationships are particularly vulnerable to disruption, and new friendships are particularly likely to form (see Chapter 9). Third, cycling females often initiate interactions with *less* familiar males whom they had shunned in the past (recall that anestrous mothers tend to avoid all males except their Friends; Chapter 5). For example, Collins (1981) reported that cycling females presented more often to males with whom they did not share a previous bond, and, in EC, cycling females groomed with a significantly wider range of males than did anestrous females (Chapter 4). These results are consistent with Rasmussen's finding that some cycling females approached a large number of different males outside of consortship. Quantitative measures showed that these females were most attractive to males, both outside of consortship and during consortship (Rasmussen, 1980)—perhaps because their approaches to many different males indicated the *absence* of strong preferences for one or two males.

With these caveats in mind, we can consider Rasmussen's results. First, although females rarely followed their consort partners, "there was a tendency for females to follow preferred male consorts more than they followed non-preferred male consorts" (1980, p. 9:27). In addition, during consortship, females groomed and presented to preferred males significantly more often than to nonpreferred males. Thus, females were indeed more cooperative when in consort with preferred males, and female cooperation appeared to affect the males' opportunities for copulation. Males mounted partners who preferred them significantly more often than they mounted partners who did not prefer them.

Although, as noted above, it is not possible to relate these findings directly to long-term affiliative relationships, they do clearly show that female preferences outside consortship are related to the behavior of both partners during consortship. It is possible that, in many cases, the long-term preferences that females show for their Friends have similar effects.

Behavior during Consort Turnovers

As noted in Chapter 7, a rival rarely attempts to take over a female in consort by approaching her directly or by interposing himself between the female and her consort partner. In fact, if a female approaches a rival, he will nearly always avoid her, even though he might have been harassing her consort partner only moments before (Collins, 1981). These and other observations described in Chapter 7 indicate that male savannah baboons are inhibited from taking a female so long as she and her consort partner maintain their normal relationship. My observations suggest that this inhibition breaks down most often in two contexts: (1) The consort male leaves the vicinity or otherwise indicates lack of interest in the female; and (2) the female breaks from her partner and runs away. The first example given below shows how the female's running away can precipitate a consort turnover. The second example shows that when the female does *not* run away, a consort turnover may not occur, even if the male is forced temporarily to abandon the female in order to respond to harassment by another male. Both examples are derived from focal samples on consort pairs during 1983, and both involve the same female, Andromeda, during peak estrus (Smuts and Watanabe, in preparation).

Dante, a young, high-ranking male, is in consort with Andromeda. Alex, Sherlock, and Boz, three older resident males, form a coalition against Dante. Over the course of 1 hr, Sherlock, Alex, and Boz charge Dante several times, and each time he wraps his arms around Andromeda and engulfs her, preventing her from moving away. Finally, the rivals charge again, and although Dante grabs Andromeda, she pulls away from him and runs off. Boz runs after her while Sherlock and Alex continue to threaten Dante. Boz catches up to Andromeda, she stops running, and a new consortship is formed (focal consort follow, 29 June 1983).

Andromeda is in consort with Alex. Dante begins to harass Alex. Zim, one of Alex's frequent coalition partners, sides with Alex, and together they chase Dante into some nearby shrubbery. We can no longer see them but hear sounds of males fighting. Several other males run to the scene and look at Andromeda. Andromeda calmly climbs a nearby hillock and sits near Sherlock, another of Alex's frequent coalition partners. They both peer into the bushes where the males are still fighting. No males approach Andromeda, and after a few minutes Alex emerges from the bushes. With Zim's help, Dante has been driven off, and Alex looks around for Andromeda. She climbs down from her vantage point and moves toward Alex, who meets her approach. Their consortship is resumed (focal consort follow, 2 July 1983).

The second example suggests that males may be inhibited from approaching a female in consortship, even if the consorting male is temporarily absent, as long as the female shows by her behavior that she has no desire for a new partner.

This inhibition, and the circumstances under which it appears to break down, resemble behaviors described in hamadryas baboons. In hamadryas baboons, males form long-term, exclusive relationships with several females, and males rarely approach or contest "ownership" of another male's females (Kummer, 1968). Experiments with captive hamadryas baboons indicate that this inhibition occurs when a rival is confronted with evidence of an existing bond between a female and another male—what Kummer refers to as the "pair gestalt" (Kummer *et al.*, 1974). The pair gestalt seems to depend, in part, on close proximity between the male and female. A hamadryas male maintains constant vigilance over the females in his group, punishing females who stray with a neck bite (Kummer, 1968). Among savannah baboons, males in consort also attempt to maintain close proximity to the female by following and herding consort partners (Hausfater, 1975; Ransom, 1981; Rasmussen, 1980). Other "possessive" behaviors, like grooming the female and mounting her in the presence of rivals, have also been interpreted as attempts by the male to reinforce the pair gestalt (Collins, 1981).

For both types of baboons, there is evidence that rival inhibition breaks down in response to female behaviors. In captive hamadryas, as noted above, males are more likely to contest ownership of females who show low preferences for their partners (Bachmann and Kummer, 1980). In savannah baboons, consort turnovers are often precipitated when the female runs away from her partner, as illustrated above. Systematic observations of consort turnovers in EC in 1983 further indicate the crucial role played by female responses. Over a 3-month period, we recorded 17 consort turnovers during peak estrus in which the behavior of the female was monitored throughout. In 14 of these (82%), the change in consort partners occurred immediately after the female broke from her partner and ran away.

In Chapter 7, I argued that harassment of consorting males by rivals might represent a form of psychological warfare designed to provoke the consorting male into aggressive retaliation that often results in loss of the female. The observations just described suggest that another function of harassment may be to provoke the female into running away. Although the rivals' threats and charges appear to be directed toward the male, the female is usually so close to her partner that she becomes a *de facto* target of harassment. As tension mounts between the males and the chances of escalated aggression increase, the female often responds by running away. Whatever the effects of her response, the immediate cause seems to be the female's fear of getting caught in the midst of a group of fighting males. In EC in 1983, 11 of the 14 female run-away responses that resulted in a consort turnover occurred just before, or just as, aggression between the consort male and his rivals erupted.

However, females do not always run away when rivals interact aggressively, as shown in the second example above. Whether or not a female runs, and when she runs, may depend in part on the identities of her partner and his rivals. It may also depend on the responses of her partner to harassment: When consorting males remain calm, females seem less prone to run away. It is possible that both the male's and the female's responses to rivals and to one another when undergoing harassment reflect their prior relationship. For example, Rasmussen (1983a) found that adolescent females, who are relatively unfamiliar with males in the consort situation, ran away from their partners significantly more often than did mature females. I had the impression that when male and female Friends were in consort, each was less vulnerable to the psychological pressures engendered by harassment. This is a hypothesis that needs systematic evaluation.

As mentioned above, many of the female run-away responses that resulted in a consort turnover occurred in the context of escalated aggression between the consort male and his rivals. However, females run away from their partners in other contexts as well, and these responses can also lead to consort turnovers. The single most common context in which a female runs away from her partner is copulation itself. During a typical copulation with a fully adult male, shortly before the male dismounts the females lowers her head, puffs out her cheeks, and, pursing her lips into a pout, she initiates a loud series of low-pitched grunts that can be heard several hundred meters away (Hamilton and Arrowood, 1978). During or immediately after copulation, the female often leaps out from under her partner and runs away, still giving the copulatory vocalization. During this "withdrawal response" (Collins, 1981), the female may run just a few steps, or, with an explosive burst of energy, she may dash up to 100 m away. The need to withdraw rapidly after copulation is apparently extremely compelling in some cases, since females show the withdrawal response even when the immediate surroundings make this awkward, or even hazardous: I once saw a female who copulated on a cliff ledge sail into the air afterward; fortunately she was only about 5 m off the ground. Another time, I saw a female who copulated on a fence post run away along the wire between two posts, in a perfect imitation of a high wire act. During the withdrawal response, females also sometimes crash into other baboons—including higher-ranking females—a *faux pas* that they would normally avoid at all costs.

Hamilton and Arrowood (1978) argue that female copulatory vocalizations advertise the consort pair's sexual activities and therefore function to incite male–male competition; the withdrawal response could serve a similar function, since it too draws attention to the fact that copulation has just occurred. These behaviors, they argue, could act to promote competition among males in general, increasing the probability that the female would end up with the highest-ranking male (see Cox and LeBoeuf, 1977, for a similar argument in elephant seals). Female responses to copulation may sometimes have this effect, especially in baboon troops like those studied by Hamilton and his colleagues in which male–male coalitions are rare and male dominance rank correlates closely with male consort activity (Busse and Hamilton, 1981). This hypothesis, however, assumes that females have no control over their responses.

Despite the apparent involuntary nature of the withdrawal response in some instances (Hall, 1962; Saayman, 1970), how far and how fast the female moves vary greatly, and sometimes she does not run away at all (Collins, 1981).

Zena has been in consort with Ovid, a young adult male, since we arrived early that morning. He copulates with her several times, and each time she shows a strong withdrawal response, running away from him 20–30 m. With the help of several allies, Alex takes Zena from Ovid, and he copulates with her soon afterward. She does not run away after the copulation and instead sits down and immediately begins to groom Alex. Ovid is following the consorting couple closely (focal sample on Zena and her consort partners, 19 June 1983).

There is also great variation in the frequency and intensity of copulatory vocalizations (Hamilton and Arrowood, 1978; Rasmussen, 1980; Collins, 1981). Collins (1981) found that the intensity of copulation calls increased significantly during peak estrus and when the male's thrusts resulted in ejaculation, suggesting that the female's response to copulation reflects the intensity of her sexual arousal. Females may also, however, have a degree of volitional control over copulatory vocalizations and the withdrawal response. If this is the case, then females could use these responses not only to incite general male–male competition but also to increase the chances that turnovers involving *particular* males will occur. Hamilton and Arrowhead state that in the chacma baboons they watched, the female run-away response was never observed to result in a consort turnover, but such turnovers occurred in EC. One example is given below.

Lysistrata, a middle-ranking female, is in consortship with Dante, a young, high-ranking male. Sherlock, Alex, and Boz, three older residents who are frequent coalition partners, follow the consort pair. The troop moves quickly through an area covered with thick bush, and Sherlock becomes separated from the others. One observer follows him, while the other stays with the consort pair. Boz and Alex continue to follow Dante and Lysistrata. During the next 2 hours, Dante repeatedly tries to mount Lysistrata, but she always refuses. The front of the troop, including Sherlock, reaches a large clearing in the bush. Sher-

lock sits on a mound at the far end of the clearing, staring intently back toward the bushes where the rest of the troop is still foraging. Gradually, baboons enter the open area, and move past Sherlock. Sherlock continues to stare toward the bushes. After ½ hr, Lysistrata and Dante, followed by Alex and Boz, enter the clearing; they are at the very back of the troop. Sherlock's body immediately tenses. Lysistrata walks into the middle of the clearing, stops, and presents her sexual swelling to Dante (this is the first time we have seen her present to Dante all day). Dante copulates with Lysistrata. The copulation lasts only a few seconds (this is typical for olive baboons), and when it is over Lysistrata pulls away from Dante and runs very quickly away from him. At exactly that moment, Sherlock rushes in from one side, and Alex and Boz charge from the other. All four males run after Lysistrata, but Boz is closest to her, and Sherlock and Alex turn to face Dante while Boz takes over the consort (focal follow on Sherlock and focal sample on Lysistrata-Dante consort pair, 17 June 1983).

While watching these events, we had the impression that Lysistrata purposefully timed her run-away response in order to increase the chances of a consort turnover. This suggestion is supported by the fact that her behavior changed from repeated rejection of Dante's advances to active solicitation at precisely the moment when a change in circumstances markedly increased the probability of a successful coalitionary challenge: The number of coalition members increased from two to three, and the animals were out in the open for the first time all morning (see Chapter 7 for further discussion of the effects of these two factors on coalitionary turnovers). Significantly, Lysistrata also appeared to dislike Dante more than did any other female seen in consort with him. She never interacted with him outside of consortship and was consistently uncooperative when he was her consort partner. A second example of a takeover from Dante by another male that appeared to be engineered by Lysistrata is given below.

Lysistrata and Dante are feeding close together. Lysistrata looks up and surveys the area. She glances at Dante, who has his back to her and is still feeding. Then she suddenly dashes off to chase a lower-ranking female feeding about 20 m away—an atypical act. During the chase, she runs directly past Sherlock, and he immediately cuts in between Lysistrata and Dante, who, after a moment of confusion, is now pursuing his partner. When Sherlock catches up to Lysistrata, she stops chasing the female and a new consortship is formed (focal sample on Lysistrata-Dante consortship, 16 June 1983).

Why Forced Copulations Do Not Occur

The observations described above indicate that female baboons do not simply acquiesce to the outcome of male–male competition. By expressing their mating preferences, females alter the costs and benefits of aggressive competition among males and thus influence the form and frequency of male–male competition. Females also influence the *outcome* of male–male competition; by merely running in the "wrong" direction when a fight breaks out, a female can influence the probability that her next consort partner will be an uninvolved bystander rather than the winner of the fight. And, finally, if the female's maneuvers fail and she does end up in consort with a male she does not prefer, she can always refuse to copulate through a simple but powerful act: sitting down.

Why, when females are only one-half the size of males, are they able to exert so much choice over sexual partners? Why do not males, who attack and even injure females in other contexts, use aggressive power to force unwilling females to mate? There are at least three possible reasons why forced copulations do not occur in wild baboons. First, as shown in Chapter 7, consort turnovers are particularly likely to occur when the female runs away from her partner, giving other males an opportunity to interpose themselves between the consorting male and the female. If a male tried to force his partner to copulate, she would probably try to escape, and if she succeeded, he would be vulnerable to a takeover attempt. Second, as we saw in Chapter 6, Friends tend to protect females from aggression by other males. If a consorting male attempted a forced copulation, the female could scream, drawing her Friend's attention to her plight. The Friend, in turn, would presumably be even more willing than usual to protect her by threatening or attacking the consort male, since these actions might provoke a consort turnover and thus give him an opportunity to mate with the female.

A third reason involves defense by females and juveniles. Females and juveniles occasionally mob males who threaten or attack adult females, particularly if the male has only recently entered the group. Although it seems likely that the female's relatives would be most willing to mob an aggressive male, unrelated animals also join in (personal observation). Mobbing by a horde of screaming females and juveniles could be costly to the male in at least three different ways. First, it would almost invariably disrupt the consortship. Second, it would attract the attention of other males, who sometimes join mobbings initiated by females and juveniles (personal observation). Finally, it could serve as an advertisement to any female in the vicinity that this was a male who abused his consort partner. As a result, these females, as well as his original victim, might avoid consorting with that male in the future.

All three of these explanations depend on the presence of conspecifics. Forced copulation has been reported in only one wild nonhuman primate, the orangutan (Mitani, 1985), the only anthropoid primate in which estrous females typically forage alone. This supports the hypothesis that the presence of other animals willing to support the female is a critical factor preventing forced copulations. This hypothesis is also supported by observations of captive chimpanzees. Forced copulations have been reported for chimpanzees kept in small cages who are isolated from friends and relatives (McGrew, 1981b; de Waal, 1982), but they do not occur in the wild or among captive animals housed in groups in large enclosures (Tutin, 1979; de Waal, 1982). In one captive group, males who persist in their attempts to mate with an unwilling female "run the risk of being chased by the female they approached and some of the other females too" (de Waal, 1982, p. 175). One female who normally sided with males in conflicts between males and females always supported the estrous female in such situations.

This discussion illustrates an important general point. Although physical strength is a significant source of power in primate social interactions, it can rarely be wielded without restraint because severe aggression usually provokes retaliation by friends or relatives of the victim. These constraints on the efficacy of physical power promote other types of competitive tactics, including alliances, psychological manipulations, and, of course, friendships.

ADULT MALE–INFANT INTERACTIONS: THE ROLE OF PATERNITY VERSUS FRIENDSHIP WITH THE MOTHER

The results reported above suggest that, as a result of female mating preferences, males who already had a bond with a female often enjoyed enhanced consort activity. This is probably one of the major benefits of friendship to males. It is also important to remember that friendships frequently involved males who were probable fathers of the female's current infant. For these males, friendships might have facilitated investment in infants already sired as well as increasing opportunities for fathering offspring in the future. In this section, I consider how probable paternity and friendship with the mother affect male–infant relationships.

Both Packer (1980) and Busse and Hamilton (1981) found that

particular types of male–infant interactions were significantly more common with males who were in the troop when the infant was conceived than with males who had transferred into the troop after conception. Packer hypothesized that frequencies of male care and protection would reflect not only male residency but also the female's consort activity during the estrous cycle in which she conceived. He also hypothesized that males who showed selectivity in consorting with particular females over time might interact more frequently with those females' infants (Packer, 1979a, 1980). Neither hypothesis was supported by Packer's results, and he concluded that, "either males were unable to assess paternity of infants . . . or the measures of consorting activity were insufficiently accurate to provide the observer with a reliable estimate of paternity" (1980, p. 515). A third possibility, however, is that the male's relationship with the infant's mother during pregnancy and lactation was a more important determinant of the frequency and nature of male–infant interactions than was the male's previous consorting activity with the mother. This hypothesis was tested in EC by comparing data on each infant's relationships with males (based on results presented in Chapter 6) with data on probable fathers and Friends of the mother. Two aspects of male–infant relationships were examined: first, affiliative interactions and, second, spatial proximity of infants to males when the mother was not nearby.

Paternity estimates were available for 13 of the infants included in the analysis of male–infant affiliative interactions described in Chapter 6. For these infants, I determined the total number of male–infant dyads falling into each of five mutually exclusive categories, defined in terms of the male's relationship with the mother and his likelihood of being closely related to the infant. For each category, I determined the percentage of dyads in which male–infant affiliative interactions were observed at least once (Table 8.4). The crucial comparison is between categories two and four: Friends who were *not* likely or possible fathers and Non-Friends who *were* likely or possible fathers. The results indicate that Friends, even when they were not probable fathers, were significantly more likely to have affiliative interactions with infants than were Non-Friend males who were probable fathers [*4*].

The second analysis considered male "associates" of infants, defined as all males found within 5 m of the infant on at least 20% of the occasions when the infant was near any male and the mother was more than 5 m away, based on Nicolson's focal samples (Chapter 6 and Appendix XIV). Paternity estimates were available for only eight of these infants (Table 8.5). Male–infant dyads were divided into five

Table 8.4. Comparison of Affiliative Interactions with Infants, Paternity, and Friendship

Type of male–infant dyad			
Male was a Friend of the infant's mother	Male was a likely or possible father of the infant[a]	Number of dyads	Percentage of dyads showing affiliative interactions[b]
1. Yes, not a son	Yes	12	83.3
2. Yes, not a son	No	10	80.0
3. Yes, putative son	No	2	100.0
4. No	Yes	12	25.0
5. No	No	192	3.0

[a]Likely and possible fathers based on consortships during conception cycles; see text for details.
[b]See text for definition of affiliative interactions; see also Table 6.4.

categories, as above, and the percentage of dyads in each category that involved male associates was determined.

Probable fathers were never associates of infants unless they also had a friendship with the infant's mother. In contrast to the findings on affiliative interactions, however, Friends were associates of infants only when they were also probable fathers. Nonetheless, this finding should be treated with caution, because the dyads included in category 2 for this analysis represent a biased sample of males. Four of six category 2 dyads involved subadult males probably born in EC; these males tended to associate only with infants who were their putative siblings (Nicolson, 1982). The other two dyads involved infants who were observed more than 5 m away from their mothers in too few of Nicolson's samples for them to have a male associate, although both infants were observed associating with the mother's Friend (SK in both cases) while away from the mother during my focal samples.

The results presented in Tables 8.4 and 8.5 suggest that: (1) For a male, the existence of a friendship with the mother is a necessary and sufficient condition for the existence of an affiliative relationship between him and the female's infant; and (2) being a probable father is likely to result in an affiliative relationship with the infant only if the male also has a friendship with the mother.

The validity of these conclusions depends, of course, on the accuracy of the behavioral estimates of paternity employed in the analysis. While their accuracy cannot be determined, the EC findings parallel the results of a study of captive macaques in which true fathers were

Table 8.5. Comparison of Spatial Proximity to Infants, Paternity, and Friendship

Type of male–infant dyad			
Male was a Friend of the infant's mother	Male was a likely or possible father of the infant[a]	Number of dyads	Percentage of dyads showing spatial proximity[b]
1. Yes, not a son	Yes	9	55.6
2. Yes, not a son	No	6	0
3. Yes, putative son	No	2	50.0
4. No	Yes	5	0
5. No	No	122	1.0

[a]Likely and possible fathers based on consortships during conception cycles; see text for details.

[b]Dyads showing spatial proximity were dyads in which the male was found within 5 m of the infant when the mother was more than 5 m away on at least 20% of all intervals in which the infant was within 5 m of any male when the mother was absent (based on focal samples by N. Nicolson; see Appendix XIV).

identified through biochemical exclusions (Berenstain *et al.*, 1981). In this study, fathers and immature offspring were found in close proximity to one another more often than other male-immature dyads, but "the effect of paternity disappeared when maternal association with males was controlled. . . . This result accords with the hypothesis that selective father-offspring association depends on the mother's relationship with males" (Berenstain *et al.*, 1981, p. 1061).

The conclusions stated above need to be confirmed in studies of wild baboons in which paternity can be determined through biochemical analysis rather than estimated by consort activity. Assuming, for the moment, that these conclusions are correct, then they imply that males sometimes formed close bonds with infants who were not their own offspring. As discussed above, one important benefit males might derive from such relationships is increased opportunity to copulate with the infant's mother when she resumes cycling. Another possible benefit is considered below.

MALE CONTACT WITH INFANTS DURING INTERACTIONS WITH OTHER MALES

A number of observers have reported that male baboons sometimes hold or carry infants during tense interactions with other males (Ransom and Ransom, 1971; Popp, 1978; Packer, 1980; Altmann, 1980; Stein, 1981; Collins, 1981, 1985; Busse and Hamilton, 1981; Busse,

1984b; Smuts, 1982; Strum, 1984) (Figure 8.1). As illustrated below, the infant's involvement in these "triadic" interactions is far from accidental. When another male is nearby, males often purposefully establish contact with a baby, usually a young, black infant only a few months old. The infant is much too small to function as an effective ally, yet its involvement often fundamentally alters the course of subsequent interactions. Why do males carry infants during encounters with other males? And why does infant-carrying so often change the outcome of an interaction? Below, each of these questions will be considered in turn.

Why Do Males Carry Infants against Other Males?

Two hypotheses have been proposed to explain why males carry infants in the presence of other males. The first hypothesis claims that males hold or carry infants to alter the outcome of an interaction with an opponent in ways that benefit the carrier (Popp, 1978; Packer, 1980; Stein, 1981, 1984; Strum, 1984). According to this hypothesis, the

Figure 8.1. An adult male takes a black infant from its mother and nuzzles it. The mother, a Friend of the male, watches calmly (she is being groomed by another baboon, obscured from view). A moment later the male carried the infant away and used it as a buffer in a tense interaction with another male.

interaction also inflicts a cost on the infant because it becomes vulnerable to injury while being carried. Thus, males "use" or "exploit" the infant (Packer, 1980; Stein, 1981, 1984). Although actual injury to infants being carried during triadic interactions has been reported in only one study (Packer, 1980), the fact that subordinate males are more likely to initiate aggression against high-ranking opponents while carrying infants (see below) and the fact that fights involving infant-carriers do occasionally occur together suggest that the risk to the infant, though perhaps small, remains appreciable (Stein, 1981).

Two types of evidence support the hypothesis that carrying an infant "buffers" a male against aggression by the opponent. First, males carry infants against males of higher rank more often than expected by chance (Popp, 1978; Packer, 1980; Stein, 1981; Busse, 1984b; Collins, 1981, 1985). Second, the outcome of the interaction may be altered in several different ways that appear to benefit the carrier (Strum, 1984). Males carrying infants are: (1) less likely to be threatened by other males (Packer, 1980; Collins, 1981); (2) less likely to retreat from a higher-ranking opponent (Stein, 1981); (3) more likely to supplant an opponent (Packer, 1980); (4) more likely to initiate aggression against higher-ranking opponents (Collins, 1981, 1985; Stein, 1981); and (5) able to resist aggression from another male for a longer period of time (Stein, 1981).

All studies agree that individual males repeatedly carry the same infant or infants over and over, and when the relationship between carriers and infants is examined, a paradox emerges: Males tend to carry the offspring of their female Friends (Table 8.6; see also Collins, 1981, 1985; Stein, 1981). This means that males carry the same infants that they care for in other contexts (Packer, 1980; Stein, 1981; Strum, 1984). It also means that males often carry infants that they have probably sired (see Table 8.6; also Stein, 1981, 1984). Why do males expose their close affiliates and probable offspring to potential danger?

Packer (1980) and Stein (1981, 1984) argue that males carry the infants they care for, rather than other infants because these infants cooperate more readily. In fact, both authors claim that males cultivate close bonds with young infants so that they will be able to use these infants in interactions against other males. Two important findings support this hypothesis (Strum, 1984): (1) Affiliated infants cooperate significantly more often than do unaffiliated infants; and (2) infant cooperation correlates positively and significantly with the carrier's ability to dominate his opponent.

According to this "exploitation hypothesis," benefits gained by an infant from a relationship with a male, such as protection from pred-

Table 8.6. Characteristics of Carriers and Opponents Involved in Male–Infant Carrying Interactions

Characteristics of males[a]	Carrier (%)	Opponent (%)	N^b
Male was a Friend of the infant's mother	96	4	45
Male was a likely or possible father of the infant	48	4	25[c]
Male was a putative son of the infant's mother	20	0	45

[a]All likely/possible fathers and putative sons were also Friends.
[b]N = number of interactions.
[c]Behavioral estimates of paternity were available for infants in only 25 of the 45 carrying interactions.

ators and other baboons (Packer, 1980) or access to better feeding sites (Stein, 1981), far outweigh the risks to the infant of being carried during interactions with other males. For the male, the costs of providing such advantages to the infant are small compared to the benefits associated with opportunities to carry the infant. Thus, it is argued, both males and infants receive a net benefit from their interactions, and the male-infant relationship exemplifies mutualism (Packer, 1980) or reciprocal altruism (Stein, 1981; Strum, 1984). According to the exploitation hypothesis, then, one major benefit to males of friendships with females involves increased opportunities to develop cooperative relationships with those females' infants; these benefits would occur regardless of the genetic relationship between the male and the infant (Stein, 1981; Strum, 1984).

Conversely, a second hypothesis stresses the genetic relatedness between the carrying male and infant (Busse and Hamilton, 1981; Busse, 1984b). This hypothesis argues that males carry their own infants in order to protect them against other males who pose a threat of infanticide (see Chapter 6). According to this theory, when a male picks up an infant who is near a recently immigrated male, he signals to the immigrant his intention to protect the infant from attack. Thus, the "protection hypothesis" supports the view that male-female friendships reflect a more general tendency for males to invest in their own infants.

In support of the "protection hypothesis," Busse found that in two baboon troops in Botswana, nearly all carriers were resident males who could have fathered the infant, and nearly all opponents were recent immigrants, who could not have fathered the infant (Busse and Hamilton, 1981; Busse, 1984b). Consistent with Busse's findings, in EC (Table 8.7) and in several other troops, residents frequently carry infants against newcomers, and newcomers very rarely carry infants at all (e.g., Packer, 1980; Collins, 1981). However, the protection hypothesis is not supported by two other findings. First, in about one-half of the 25 interactions in EC that involved infants conceived after the study began, the carrier had never been seen consorting with the mother and was therefore not a likely father of the infant (Table 8.6). Second, although Stein (1981) reported that probable fathers (males who had consorted with the mother during her conception cycle) carried infants more often than expected by chance, opponents were also probable fathers more often than expected, and the male carrying the infant was more likely to be its father than the opponent in only 42% of the interactions. Packer and Pusey (1985) reported almost identical results: The carrier was more likely than his opponent to be the father only 40% of the time.

Despite a great deal of information about male carrying of infants, neither hypothesis has yet been falsified. One reason the debate remains unresolved is that these two hypotheses share many predictions (Busse, 1984b; Collins, 1985). For example, both predict, for different reasons, that males will carry infants primarily against higher-ranking opponents and that males will carry infants with whom they have caretaking relationships. But I think that there is a more fundamental reason why neither hypothesis has been falsified: Both are probably correct. Most authors have assumed that male carrying of infants—at least in a given troop—can be explained by one hypothesis or the other. However, there is no *a priori* reason to assume that triadic interactions represent a unitary phenomenon requiring a single explanation (Smuts, 1982; Collins, 1985). A number of observations suggest just the opposite: Within a troop, male carrying of infants may range from pure protection to pure exploitation, and everything in between. These observations are summarized below:

1. *Some carrying interactions appeared to be protective while others seemed to be exploitative.* In EC, six distinct types of triadic interactions occurred, distinguished by the context of the interaction and by the behavior of the participants (Table 8.8). Roughly one-third of these interactions were consistent with the protection hypothesis; another

Table 8.7. Percentage of Male–Infant Carrying Interactions Involving Different Combinations of Carriers and Opponents, by Age/Residence Status[a]

	Opponent				
Carrier	Long-term resident	Short-term resident	Newcomer	Subadult or young adult natal male	Total
Long-term resident	8.2	12.2	10.2	4.1	34.7
Short-term resident	10.2	4.1	14.3	6.1	34.7
Newcomer	0	0	0	0	0
Subadult or young adult natal male	12.2	10.2	4.1	4.1	30.6
Total	30.6	26.5	28.6	14.3	100.0

[a]This table is based on 45 carrying interactions. In three interactions, there were 2 opponents, and in one, 3 opponents, for a total of 49 carrier/opponent dyads. In only 1 of the 45 interactions was an opponent not in the troop when the infant was conceived. This reflects the fact that no males transferred into the troop during the study.

one-third were more consistent with the exploitation hypothesis, and a final one-third were too ambiguous to classify. Examples of protective, exploitative, and ambiguous interactions follow.

(Protection) *Jocasta's 11-month-old infant, Jason, is feeding away from his mother. Triton, a recent immigrant, moves past Jason. Jason looks at Triton, fear grins, and gecks. Hector, an older resident male who is a Friend of Jocasta, is travelling about 5 m away. As soon as he hears Jason geck, he runs toward him. At the same instant, Jason runs away from Triton and into Hector's arms. Hector sits holding the infant and stares at Triton. Triton ignores the interaction and continues walking (focal sample on Jocasta, 15 August 1978). (This interaction was scored as both Type 1 and 3, since both the male and the infant initiated contact.)*

(Protection) *Boz, an older resident, is feeding. Homer, a natal adult male, is resting about 20 m away. When Homer sees his*

Table 8.8. Different Types of Male–Infant Interactions Observed in EC[a]

Type of interaction	Percentage of all interactions
Protective	
1 An infant shows distress in response to an approach or proximity of another male. The carrier rushes to the infant in response to its vocalizations and either holds or carries it.	7.5
2 The male sees the infant near or moving toward another male and intercepts the infant, holding or carrying it and thereby preventing it from getting closer to the other male.	17.5
3 The infant initiates the carrying interaction by approaching a male who is near one or more other males.	7.5
All protective interactions	32.5
Exploitative	
4 A male approaches an infant and then carries it toward another male, initiating or continuing an agonistic interaction.	30.0
All exploitative interactions	30.0
Ambiguous	
5 The male is already in close proximity to the infant. Another male moves toward them, and the first male grabs the infant. He may then carry the infant or simply hold it while watching the other male.	32.5
6 The male approaches and holds or carries an infant immediately *after* an agonistic interaction with another male has ended.	5.0
All ambiguous interactions	37.5
All interactions	100.0

[a]Based on 39 male–infant carrying interactions (in 6 others, the start of the interaction was not observed, and so it was not possible to classify it). One interaction was classified as both Type 1 and Type 3 (see example in text), for a total $N = 40$.

infant sister, Ligeia, moving toward Boz, he runs to her and moves in between her and Boz. He carries Ligeia a few feet away and then grooms her. Boz ignores Homer (ad lib observation, 10 August 1979). (Type 2 interaction.)

(Exploitation) *Boz, his Friend Phaedra, and her 2-month-old daughter, Pasithea, are feeding in close proximity. Triton, a recent immigrant, also feeds nearby. Boz pulls Pasithea ventral and approaches Triton. Triton lunges toward Boz, who calmly walks away. Triton follows Boz and repeatedly yawns to display his canines. Phaedra runs a few meters away to avoid Triton and watches. Boz sits facing Triton with Pasithea ventral. Triton continues to yawn at him, but Boz remains calm. Finally Triton backs off, and Pasithea returns to her mother. Boz follows them and all three adults resume feeding (ad lib observation, 23 August 1978). (Type 4 interaction.)*

(Exploitation) *Triton, a recent immigrant, is in possession of an adult impala carcass appropriated from a group of jackals. After feeding for a few minutes, Triton moves a few meters away to rest, maintaining a careful guard over the meat. Cyclops, an older resident, moves to within about 2 m of the carcass, opposite Triton, and glances at the other male. Triton immediately threatens Cyclops, returns to the meat, and resumes feeding. Cyclops moves a few meters away and looks toward his Friend, Phoebe, and her infant, Phyllis, who are resting nearby. He grunts at Phyllis, and she runs over and leaps onto his back. Carrying Phyllis, Cyclops strides confidently toward Triton and sits down opposite him, with the carcass between them. Phyllis climbs off Cyclops' back and plays nearby. Triton immediately stops feeding and yawns at Cyclops, exposing his fine canines. Cyclops gathers Phyllis into his lap and lip-smacks and grunts at her. Triton continues yawning. The two males sit about 1 m apart for several minutes. Neither feeds, and they avoid looking at one another: It is a stand-off. Then Phyllis, who has been sitting quietly in Cyclops' lap, crawls onto the carcass and begins to bounce up and down in play; she is exactly midway in between the two males. Cyclops grunts at Phyllis and immediately begins to feed. At the same moment, Triton abandons his position and leaves the area (ad lib observations, 25 August 1978). (Type 4 interaction.)*

(Ambiguous) *Handel is resting with his Friend Psyche and her 6-month-old infant, Penelope. Homer, a natal male, walks by about 10 m away and glances at them. Handel pulls Penelope ventral and, simultaneously, threatens and pant grunts at Homer. Homer moves away quickly. Handel lets go of Penelope. Psyche ignores the interaction (focal sample on Psyche, 17 November 1978). (Type 5 interaction.)*

(Ambiguous) *Adonis, a natal male, is drinking from a puddle while Midas, his 2-month-old brother, watches. Boz, a higher-ranking resident male, approaches as if to supplant Adonis. Adonis grabs Midas, pulls him ventral, and moves away from the puddle. Boz drinks while Adonis and Midas watch (focal sample on Medea, Midas's mother, 6 June 1978). (Type 5 interaction.)*

In the last two examples, it is difficult to tell whether the male held the infant because he viewed the proximity of the other male as a threat to the infant, or in order to highlight a threat (first example), or exit more gracefully from a tense situation (second example). I favor the second explanation in each case, but without knowledge of the motivations of the males, these interpretations are simply educated guesses.

2. *The behavior of infants and mothers during carrying interactions varied from cooperative to resistant.* The second example of exploitation, above, illustrates active cooperation by an infant. Occasionally, infants actually initiate carrying events by running to an affiliated male who is interacting with another male even before the first male solicits the infant's aid (Ransom and Ransom, 1971; Stein, 1981). As the Ransoms pointed out (Ransom and Ransom, 1971), triadic interactions in which infants play an active role resemble the coalitions formed by adult males (see Chapter 7)—except that in this case, one of the allies is conspicuously small.

More often, infants play a less active, but nonetheless cooperative, role, and several researchers have emphasized the fact that infant cooperation facilitates male use of infants during interactions with other males; however, infants and their mothers do not always cooperate with male carrying attempts (Altmann, 1980; Stein, 1981; Nicolson, 1982; Strum, 1984). During most carrying events in EC, the mothers ignored the interactions, but in 3 of 45 cases, the mother tried to retrieve the infant from the carrying male, and in 6 cases the infant screamed while being carried (3 of these were the same events in which the mother resisted the male). The beginning of 2 of the interactions was not observed, but the other 4 could be categorized: Two were Type 4 interactions (exploitative) and 2 were Type 5 (ambiguous) (see Table 8.8). Thus, mothers and/or infants were never observed to express distress during those events that were consistent with the protection hypothesis (Types 1–3). Furthermore, in only 2 of the 45 events was the carrier not a Friend of the infant's mother, and in 1 of these cases both the mother and the infant actively resisted the carrier. Such resistance was not seen in any of the other events.

These results suggest that a detailed analysis of the responses of mothers and infants in a large sample of triadic interactions (including those in which the male's attempts to maintain contact with the infant fail), might help to clarify which situations involve primarily protection and which exploitation. It would be important in such an analysis to consider mothers' and infants' responses in terms of the behavior of both the carrier and opponent and also in terms off the males' identities (e.g., Friend of the mother, resident, or newcomer male).

3. *Males exploited and protected infants against both residents and immigrants, and males both exploited and protected the same infant.* Whether a carrying interaction is exploitative or protective appears not to depend on the identity of the opponent or the carrier. Some of the variation in triadic interactions observed within a single troop might possibly relate to variables that have been stressed by proponents of both the exploitation and protection hypotheses. On the one hand, we might expect that males would show protection primarily against recent immigrants, who are least likely to have fathered infants and who therefore represent the greatest threat to them. On the other hand, when males carry infants against residents, we might expect to see them using infants to protect themselves from aggression, since a resident should be reluctant to initiate aggression toward a male carrying an infant to whom he, the resident, might be related. These hypotheses, however, are not supported by the data given in Table 8.9, which show that the proportions of residents, immigrants, and newcomers that were the targets of protective versus exploitative interactions were roughly equal. Another possibility is that the degree to which triadic interactions are exploitative or protective varies among male–infant dyads. I do not have enough observations of the same dyads interacting repeatedly to evaluate this hypothesis quantitatively, but in several cases the same male was seen both to protect and to exploit the same infant.

> (Infant Protection) *Triton is feeding. Psyche is feeding 20 m away with her 6-month-old infant nearby. Handel, Psyche's Friend, feeds near Psyche. Psyche's infant, Penelope, moves to within 3 m of Triton and shows "ambivalence" (Packer, 1979a): She shifts toward and away from him repeatedly, and peers at him while gecking in fear. Handel runs in between Penelope and Triton, sits down in front of the infant, and pulls her into his*

Table 8.9. Percentage of "Protective" and "Exploitative" Male–Infant Carrying Interactions Involving Opponents of Different Age/Residence Status

Opponent's age/residence status	Protective[a] (*N*=13)	Exploitative[b] (*N*=12)
Long-term resident	31	25
Short-term resident	15	25
Newcomer	46	42
Natal	8	8
Total	100	100

[a]Protective interactions included Types 1, 2, and 3 in Table 8.8.
[b]Exploitative interactions included Type 4 in Table 8.8.

lap. Triton watches and then resumes feeding. Handel carries the infant back to Psyche and resumes feeding (ad lib observation, 9 October 1978). (Type 2 interaction.)

(Exploitation) *Handel and Alex are feeding about 20 m apart. Over the past few days, Alex has repeatedly attempted to initiate greetings with Handel, who always avoids his approaches.*[4] *Handel seems very tense, and he darts quick glances at Alex whenever he moves. After a few minutes, Handel wanders close to Alex while foraging but avoids looking at his rival. Psyche, Handel's Friend, follows him, and her 7-month-old infant leaps off her back onto the ground, catching Handel's attention. He glances at Alex, who is still feeding, and then steps toward Penelope, pulls her ventral, and sits down right in front of Alex. Handel lip-smacks noisily and grunts at Penelope while glancing at Alex. Alex, without looking at Handel, turns and leaves the area. Handel, Psyche, and the infant resume feeding (ad lib observation, 27 November 1978). (Type 4 interaction.)*

4. *Many carrying interactions combined elements of both protection and exploitation.* In some cases, what began as protection seemed to change to exploitative carrying as the interaction progressed.

[4] Male greetings involve a rapid exchange of "sexual" gestures including perineal present, grasping of the hips, mounting, and sometimes touching of the genitals. These encounters are brief and usually very tense. Their function is not well understood, but they may play a role in establishing alliances between males (Smuts and Watanabe, in preparation).

Homer notices his sister, Ligeia, playing near Triton, a recent immigrant. Homer runs to Ligeia, pulls her ventral, moves a few meters away from Triton, and sits. Triton ignores Homer. Homer watches Triton for a moment and then carries Ligeia toward him while exposing his canines in threat. Triton moves away (ad lib observation, 27 July 1979). (Type 4 interaction initially; Type 2 near the end.)

In other cases (category 5 in Table 8.8), the interactions were ambiguous from the start, as shown by the two examples of ambiguous interactions given earlier.

In summary, male carrying of infants can be both protective and exploitative, and it is often difficult to determine which component predominates in any given interaction. Males carry the same infants in both protective and exploitative ways, and the infants they carry are nearly always the offspring of their female Friends—the infants they care for in other contexts. These results indicate that protection and exploitation of infants are intimately linked, and there is no simple answer to the question of why males carry infants against other males (see also Collins, 1985). Sometimes males carry infants to protect them; at other times males carry infants to protect themselves; and sometimes males appear both to protect and exploit infants in the same interaction.

Why Does Infant-Carrying Inhibit Aggression?

Although no consensus exists concerning why males carry infants, most observers agree that carrying an infant makes a male less vulnerable to aggression (Packer, 1980; Stein, 1981; Collins, 1981; Strum, 1984). Why should this be so?

The protection hypothesis suggests one possible answer to this question. Evolutionary theory predicts that when the payoff to winning a fight varies greatly between two opponents that are fairly evenly matched, the contestant that has the least to gain should defer to the other animal, and the contest will be settled without a fight (Maynard Smith and Parker, 1976). Busse (1984b) has used this reasoning to explain why males protecting infants are rarely attacked by rivals: The payoff to the carrying male, involving survival of an infant likely to be his own offspring, is greater than the payoff to the opponent, consisting of an earlier opportunity to mate with the infant's mother. Thus, an opponent who attacks a male who is carrying his own infant is likely to be the target of fierce retaliatory aggression by the carrier.

Collins (1981, 1985) and Packer and Pusey (1985) take this argument one step further, proposing that the prior inhibition against attacking a male carrying his own infant can be manipulated by males who use infants that may or may not be their own as agonistic buffers in a wide variety of situations. Thus, use of infants as buffers would be exploitative in two senses: The carrier exploits the infant by exposing it to risk of injury, and he also exploits the opponent's inhibition against attacking an infant that is being protected by a likely father.

If this hypothesis is correct, then it can be argued further that opponents would benefit from an ability to distinguish true protective carrying—a situation in which aggression would be very risky—from exploitative carrying—a situation in which it might sometimes pay to "call the bluff" of the carrier. Carriers could, in turn, make it more difficult for opponents to distinguish between protection and exploitation by combining elements of both in their behavior toward infants and opponents. If so, the ambiguous behaviors that make it difficult for observers to interpret triadic interactions may reflect the carrier's attempts to confuse his opponent. In other words, a carrier may use protective behavior to deceive opponents about his motives for carrying the infant, thereby reducing his vulnerability to aggression.

This scenario is consistent with recent theoretical attempts to understand animal communication in terms of the individual costs and benefits associated with the interaction (Popp and DeVore, 1979; Dawkins and Krebs, 1978; Caryl, 1979; Cheney and Seyfarth, 1985; Trivers, 1985). Natural selection, it is argued, will favor continuous improvements in individual abilities to manipulate the behavior of others in ways that increase reproductive success. Natural selection will also favor continuous improvements in individual abilities to detect and thwart manipulation by others.[5] According to Dawkins and Krebs, "the general conclusion is that bluff and deceit are always advantageous, but they are limited by probing and assessment" (1978, p. 304).

It is important to stress that this evolutionary model does not imply a conscious intention to manipulate or to detect deception on the part of the animals. Terms like *deceit* and *manipulation* are regarded as convenient labels for behaviors that, it is argued, have been favored by selection because of their effects on individual reproductive success.

[5] The evolutionary process proposed here is similar in principle to continuous improvement in the hunting capacities of predators and in the prey's capacities to elude capture, except that in this instance the antagonists are members of the same species, and they will find themselves in the roles of both manipulator and victim over the course of many interactions.

However, in animals as intelligent as baboons, it is possible that a tendency to deceive or to manipulate opponents during triadic interactions results partly from learning (Collins, 1981, 1985), and in some cases the ambiguous behaviors of carriers may reflect a conscious attempt to confuse the opponent.[6] Alternatively, the carrier may experience conflicting emotions, and his equivocal behavior may simply reflect his own ambivalence. Such ambivalence could provide a mechanism for manipulation of opponents, since self-deception concerning one's own motives may sometimes be the most effective means of deceiving others (Trivers, 1985).

In some cases, exploitation of an inhibition against attacking infants who are being protected by a likely father may help to explain the reluctance of males to show aggression toward rivals carrying infants. However, it remains inadequate as a general explanation for a reduction in the opponent's aggression. For example, it does not explain why opponents fail to attack rivals when the infant is clearly being exploited (e.g., when a male runs to pick up an infant after another male initiates aggression against him) or in situations in which the carrier cannot effectively retaliate should the opponent attack (e.g., when a juvenile male uses an infant against a fully adult male). Why are males generally reluctant to attack rivals carrying infants in a wide variety of situations?

The most plausible general explanation rests not on the behavior of the carrying male in the event of an attack but on the behavior of other troop members. The involvement of the infant in the interaction results in a sharp increase in the number of baboons who have a stake in the outcome; it becomes a public event, and a male who attacks an infant is subject to mass retaliation, particularly by the infant's relatives and close associates (Smuts [cited in Hrdy, 1979], 1982; Stein, 1981; Strum, 1984). Mobbing of males in this context has been observed in at least six different troops (Mountain Zebra Park: Cheney, 1977b; Amboseli: Stein, 1981; Ruaha: Collins, 1981; Moremi: Collins *et al.*, 1984; PHG troop at Gilgil: Strum, 1984; EC: personal observation). During mobbing, a horde of screaming females and juveniles (and sometimes other males) descends upon the offending male with such speed and ferocity that he flees instantly. Sometimes mobs chase males several hundred meters beyond the periphery of the troop. It is difficult to convey just how terrifying mobbing is. The first time I witnessed it, I was extremely alarmed even though I was quite far away, and I once saw an

[6] See Menzel (1979), Woodruff and Premack (1979), and de Waal (1982) for evidence of intentional deception in captive chimpanzees.

uninitiated human observer turn and run when a group of females and juveniles mobbed a nearby male, even though I had just warned him that a mobbing was about to occur.

Being mobbed could be disadvantageous to the opponent for two reasons. First, he might be injured, especially if other males joined in. Second, mobbing, or counterattacks by other males, could serve as an "advertisement" to all females in the troop that the attacking male is a source of danger to infants; if females are reluctant to form friendships with such males and to mate with them, advertisements of this type could reduce a male's future reproductive success. Stein (1981) presents a similar argument.

Having access to cooperative infants thus may confer important advantages to male Friends in interactions with other males. The benefits of friendship to males include not only long-term reproductive advantages but also short-term social advantages resulting from manipulation of associations with infants, whether theirs or someone else's.

SUMMARY AND CONCLUSIONS

Males receive several important benefits from their female Friends. First, a prior bond with an anestrous female significantly increases a male's chances of forming consortships with this female in the future when she resumes sexual cycles. Second, males are able to use both female Friends (Chapter 6) and those females' infants (this chapter) as buffers during agonistic encounters with other males. Third, in many cases but by no means all, the male's relationship with the mother provides an opportunity for investment in his own likely offspring; this may include protecting the infant against aggression from other males during carrying interactions and in other contexts. A fourth possible benefit of friendships with females involves integration of immigrant males into a new group; this will be discussed in Chapter 9.

It is impossible to say which of these benefits is most important to males, in part because the data are not yet adequate to the task, but also in part because the importance of each benefit probably varies depending on the male and on his relationship with the female. For short-term resident males, the most important benefits of friendship are probably acceptance into the group (Chapter 9) and increased opportunities to mate, rather than opportunities for paternal investment. This suggestion is supported by two observations. First, young, short-term residents often formed friendships with females whose infants they were very unlikely to have sired. Second, most (61%) of their consort partners were Friends. Older long-term residents, in contrast, generally restricted their friendships to females with whom they had consorted during conception cycles, indicating that opportunities for paternal investment might be an important benefit of friendship for these older males. Older males also benefit from increased mating opportunities with Friends. However, because they are familiar to all troop females, they may rely less on friendship as a means of demonstrating their good qualities as potential mates than do recent immigrants. This suggestion is supported by the fact that 72% of the consort partners of long-term resident males were Non-Friends.

However, both types of males gain benefits related to use of female Friends and those females' infants during agonistic encounters with other males. Chapter 6 showed that although males protect their female Friends and their offspring against male aggression, being Friends with a male also exposes a female to use as an agonistic buffer and to attacks by males attempting to challenge her Friend. This chapter makes it evident that infants, too, are sometimes exposed to dangerous situations when interacting with their male associates. From the point of view of females and infants, then, friendships with males appear to represent a trade-off that is far from ideal: The female and infant gain the protection of a powerful ally, but they are, in turn, made more vulnerable to particular forms of exploitation. I suggest that natural selection has favored females and infants willing to accept this trade-off, because forming a friendship with one or two particular males is the best way for a female and her infant (1) to avoid harassment by higher-ranking females and juveniles and (2) to reduce the threat of violence by males in pursuit of reproductive opportunities.

It should be clear by now, however, that male baboons do not have all the power in male-female relationships. Although the male's large size and superior fighting ability clearly affect the nature of male-female relationships, female baboons ultimately determine whether a male becomes an accepted member of a troop (see Chapter 9), and as we have seen here, once a male enters a troop, female preferences strongly influence his mating success. Females may use their control over benefits that males cannot appropriate by force to acquire concessions from males. For example, one reasonable interpretation of the relationship between male–female friendship and female mating preferences observed in EC is that females prefer to mate with males who have already demonstrated a willingness and ability to invest in them and in their offspring. If enough females show such preferences, males may be forced to modify their reproductive strategies in ways that benefit particular females (cf. Tutin, 1979).

These ideas are very speculative in terms of the specific costs and benefits discussed. However, the results reported in this book provide strong support for the more general conclusion that friendships in baboons reflect mutually advantageous compromises between the often conflicting reproductive interests of females and males. Because their interests are not identical, and because opportunities for alternative arrangements are always present, these compromises inevitably reflect dynamic processes rather than stable solutions. From this perspective, it makes sense to view the social interactions associated with the development, growth, and dissolution of friendships as behavioral mechanisms that allow baboons to continuously negotiate the terms of the relationship. These interactions are the subject of Chapter 9.

NOTES ON STATISTICS

[1] Comparison of the number of male–female dyads that consorted more often than expected for Friends versus Non-Friends, based on 19 estrous cycles of 8 females who conceived early in the study (8 conception cycles and 11 nonconception cycles totalling 72 consort-days). $\chi^2 = 4.3$, $d.f. = 1$, $N = 80$, $p < .05$.

[2] Comparison of the number of likely fathers that were young adult, short-term residents or natal males versus those that were older, long-term resident males early and late in the study. Fisher exact probability test, $N = 23$, $p = .007$.

[3] Comparison of the number of male–female dyads that consorted more often than expected for Friends versus Non-Friends, based on 33 estrous cycles of 14 females who conceived late in the study or after the study ended (8 conception cycles and 25 nonconception cycles totalling 145 consort-days.) $\chi^2 = 7.2$, $d.f. = 1$, $N = 36$, $p < .01$.

[4] Comparison of the number of male–infant dyads seen to have affiliative interactions for males who were Friends of the mother but who were not probable fathers of the infant versus males who were not Friends of the mother but who were probable fathers of the infant. Fisher exact probability test, $N = 22$, $p = .015$.

9 MAKING, KEEPING, AND LOSING FRIENDS

Hector responds with polite interest to Delphi's presentation of her hindquarters for Hector's inspection. This is not always a sexual gesture, since females present at all stages of their reproductive cycles. Responsiveness to gestures like these can be used to communicate a desire for a closer relationship; failure to respond can indicate a disinterest in pursuing the relationship.

INTRODUCTION

A number of intriguing questions about baboon friendships have been left unanswered by the previous chapters. How are friendships formed? Why do they sometimes end? How do they vary between pairs? How long do they last? What emotions are associated with making—and losing—friends? These are among the most basic questions one can ask about any important relationship; they are also among the most difficult to answer. This difficulty exists partly because revealing events like the start or end of a friendship are uncommon, and partly because collection of data about the quality of a relationship, or the subtle dynamics of interactions, requires a sensitivity to baboons and an attention to detail that come only after a long period of data collection on more straightforward topics.

For these reasons, the questions posed above can only be addressed in an anecdotal fashion. In this chapter, some of my impressions (and those of others when available) about friendships between female and male baboons are described. Whenever possible, these impressions are supported with examples or data. I hope that the ideas presented below will at least help to satisfy curiosity about some of the more elusive aspects of baboon friendships and, at best, stimulate further interest in the behavioral mechanisms animals use to create long-lasting relationships.

BABOON ADOLESCENCE: MAKING SEXUAL FRIENDSHIPS FOR THE FIRST TIME

Immature females sometimes have special relationships with one or two males. They spend time near these males and groom them, and the males, in turn, protect the females from other baboons and allow them to feed nearby. In most cases, these relationships are based on bonds the females developed during infancy with their mothers' Friends (Stein, 1981; Johnson, 1984). In EC in 1983, there were several pairs of older adult males and adolescent females who had shared a close bond when the female was an infant, and these animals continued to show preferential association. I never observed any sexual interactions between them, although these females frequently copulated and consorted with other males. These observations corroborate reports that older males present in the troop when an adolescent female was conceived rarely show sexual interest in that female (Packer, 1979a; Scott, 1984). Thus, it seems likely that baboons experience an inhibition against mating between possible fathers and daughters, particularly if the animals have shared a close bond while the female was growing up. Whether these bonds ever persist after the female becomes an adult is not known. More often than not, the males that a female was close to as a youngster will have died or left the troop by the time she has had her first infant.

With the exception of these familiar "father-figures," immature females generally avoid adult males, and they never have sexual interactions with them. Like their mothers, they are particularly nervous around young, recent immigrants (Packer, 1979a).

All of this begins to change when the female reaches menarche, between 4 and 5 years of age. At first, her diminutive appearance and playful antics resemble more closely the juvenile period that is ending than the adult sexuality and motherhood that are to come. A pubescent female is little more then one-half the weight of a mature female, and she will not achieve full adult size until a few years after she first conceives (Scott, 1984). For the first year or so after menarche, adolescent females are indistinguishable from their juvenile playmates, except for the telltale pink swelling that begins to blossom a few days after each menses.

During the first few cycles, the perineal swelling is quite small, and most of an adolescent female's sexual interactions involve playful mounts by immature males. Soon, however, she begins to show intense sexual interest in adult males. Initially, this desire is one-sided, but after she undergoes six or seven cycles, adult males begin to return her interest (Scott, 1984). This period seems to be one of great social and emotional turmoil for the female. Suddenly, she finds herself attracted by, and attractive to, individuals that she has always avoided. Her interactions with males reflect this tension between sexual desire and fear. At the same time, a playful element often intrudes, as if she cannot quite bring herself to take her emerging sexuality seriously.

The baboons are feeding in an open field. Euphoria, a slight adolescent female with a small, shiny swelling, surveys the field, and then approaches Dante, a young, high-ranking male. When she is about 3 m away from him, she swivels her body around and then begins slowly to back up to him, glancing nervously over her shoulder all the while. At one point, Dante looks up, and Euphoria jumps away, but he does not look at her. She resumes backing toward him. It takes her 2 minutes to come within touching distance, and once there, she holds her rear up as high as she can, looking back at Dante expectantly. Dante ignores her and continues to feed. After a minute or so, Euphoria bounds off and repeats her prolonged, backing-up approach with another adult young male, Orpheus. Orpheus ignores her as well. Euphoria takes a few moments out to play with a black infant, and then looks around for another male. She spots Boz, an older resident, who is resting under a tree. Again, she slowly backs up, and again, she lifts her rear as high as she can and peers back at Boz. Boz, a particularly large male, continues to stare straight ahead, apparently lost in a baboon daydream. Because Euphoria is so small, her swelling is not directly within Boz's line of sight; it is instead on a level with the lower part of his torso. Euphoria seems to realize this disadvantage, and she leans forward on her elbows, slowly lowers her head to the ground, raises her rump high into the air, and peers back at Boz upside down from between her legs. Even the imperturbable Boz cannot resist an adolescent female standing on her head, and he grunts, lip-smacks, and leans forward slightly to sniff her swelling. Euphoria leaps away with a cry, and then turns to look at Boz from a few feet away, her tail up in alarm and her mouth curved into a fear-grin. Boz grunts reassuringly, and Euphoria moves off to try her new approach on another male (ad lib observation, 16 June 1983).

It is during this tumultuous period of ambivalence and high sexual energy that a female baboon begins to develop her first sexual friendships with adult males. No published information exists on how these relationships develop, and my own observations are meager. I have the impression that many adolescent friendships take a long time to develop, because the females' misgivings about unfamiliar males must be overcome gradually. Direct approaches by potential Friends often frighten adolescent females, and more subtle initiatives, which allow the female to "set the pace," seem to be more effective. (For an example of this type of approach, see the account of Thalia and Alexander in the Prologue.)

In EC, the relationship between friendship and mating was particularly strong for adolescent females. Four adolescent females (AM, AU, DL, and PH) consorted almost exclusively with their Friends (23 of 25 days in consort). Another, TH, consorted exclusively with her Friend, AA, and a subadult male, PL, until her conception cycle, when two older residents monopolized her. This close correspondence between friendship and consort partners perhaps reflects a particularly strong desire to consort with Friends or to avoid consorting with unfamiliar males. However, it may also result from reduced competition among males for consortships with adolescent females, who are less likely to conceive during an average cycle than are adult females (Scott, 1984) and less likely to raise an offspring to maturity if they do conceive (Nicolson, 1982).

As shown in Chapter 4, young females tend to form friendships with young males. Five different males had friendships with adolescent females. Four of the five were young adults or subadults, and only one (Boz) was a resident male (Boz appeared to be the youngest of the five resident males). Young adult and subadult natal males showed a strong tendency to consort mainly with adolescent females (23 out of 29 consort days).

After giving birth, primiparous females in EC sometimes formed new associations with subadult natal males who were not frequent consort partners. Ransom (1981) reported that among Gombe baboons, young males (approximately 4–10 years old) tended to form strong bonds with the infants of young females. At Gombe, the subadult male and the infant's mother "showed no more intense relationship than mutual tolerance" (p. 229), but in EC these males tended also to develop a friendship with the mother. However, the relationship between the male and the mother appeared to originate from the male's interest in the infant, rather than the reverse, which is more typical of older pairs. In EC, as at Gombe (Ransom, 1981), these relationships always involved females of low rank.

Subadult and young adult males also interacted frequently with their putative siblings; examples include the relationship between Pliny and Daphne's infant and between Adonis and Medea's infant. It seems likely that through their interactions with their siblings and the infants of young, low-ranking females, subadult males learn social skills that will be of use to them when they later transfer into another troop and attempt to form relationships with unfamiliar females and infants.

Ransom (1981) had the impression that young females (and, by implication, young males) had weaker friendships than did older pairs. In general, I share this impression, but I found some notable exceptions. One of the most intense bonds I have ever seen in baboons was between Cicily and Pegasus (see Chapter 6 for a description of Pegasus' defense of Cicily). Scott (1984) also refers to a particularly intense bond between an adolescent female and a subadult male.

MALE IMMIGRANTS: MAKING FRIENDS WITH STRANGERS

Entering a New Group

There is evidence that both female and male baboons are attracted to opposite-sexed individuals from other troops (Packer, 1979a). Before transferring, males spend a lot of time monitoring other groups, and during intergroup encounters they sometimes succeed in copulating with strange females (Cheney and Seyfarth, 1977; Packer, 1979a; Smuts, personal observation). These interactions seem to influence male transfer: In one troop, four subadult males emigrated after an intergroup encounter that included copulations, and they left in the company of cycling females from the other group (Cheney and Seyfarth, 1977, p. 404). The example below provides a vivid illustration of female attraction to a strange male and of the role females can play in enticing new males into their troop. I saw this event in 1976 during a pilot study of a troop of olive baboons in Masai Mara Game Reserve, Kenya.

The study troop, KF, and another troop, GD, spent the night in the same sleeping grove, a few hundred meters apart. KF, the subordinate troop, left the trees early and moved about 500 m out on to the plains, where they fed. GD animals rested and fed along the edge of the sleeping grove. Chris, a young KF female with a full sexual swelling, sat on the edge of KF staring at GD. After several minutes of intent scanning, she stood up abruptly and walked quickly across the gap between the two troops. Slowing her pace, she sauntered into GD and moved directly toward Aaron, a large subadult male. When she was a few meters away, she turned and began backing up, glancing at him repeatedly over her shoulder. Aaron, who was feeding, looked up and then all around, as if wondering who was the target of this strange female's attention. Chris continued to move slowly closer, and Aaron's attention became riveted on her swollen bottom. When she was about 2 m away, he made the "come-hither" face accompanied by a chin pull—a response that invites the female to present her rear for the male's tactile inspection. Chris, glancing over her shoulder, saw Aaron's expression and immediately bounded away. Then she stopped, looking back at him. Aaron resumed feeding, but he kept stealing glances in her direction. She slowly backed toward him. Once again he made the "come-hither" face, and once again she bounded away. This time he followed her for a few meters and then sat. She backed up, getting closer this time. He reached for her hips, lip-smacking. She leapt away. This sequence was repeated several times.

Chris's timing was flawless, and she soon had Aaron so engrossed in his attempts to get close enough to touch her that he became oblivious of all else. She led him out of his troop and into the area that separated GD and KF. By the time they had nearly crossed the gap, her coy manipulations had reduced Aaron to a caricature of a male baboon responding to a present by an attractive female: He sat with arms outstretched, both hands in the position they would have been in had he been grasping Chris's hips, lip-smacking, making the "come-hither" face, and alternately glancing at the empty space in front of him where Chris's rump should have been, and at Chris, calmly watching him from 2 m away. Chris seemed to conclude that this was the moment for decisive action. She rushed toward Aaron, swerved away as she passed within centimeters of him, and ran directly into her own troop. The ploy worked. Aaron ran after her, seemingly unaware of KF until he was only a few meters away from several KF baboons. He skidded to a halt, gazed around at KF, and then back at GD for the first time since Chris had coaxed him away. Then he ran into KF and chased several females. Chris approached and began to groom him, but she was soon supplanted by a higher-ranking, cycling female.

From that moment on, Aaron was a member of KF troop.[1] He was still there 5 years later, by which time he had become the most successful consorter in the troop (Sapolsky, 1983; Aaron is referred to as no. 257). Aaron might have transferred into KF eventually even if Chris had not lured him away from GD, but there is no doubt that on that day, Aaron found himself in a new troop as a direct result of Chris's deliberate and skillful efforts.

[1]As far as I knew, Aaron had never been in KF troop before, and he had certainly not been a member of the troop during the preceding 4 months. Although Aaron copulated with Chris after he entered her troop, they did not form a friendship during the next few months, perhaps because she already had a close bond with a prime adult male.

Being Accepted

In addition to influencing the timing of male transfer, females also strongly influence an immigrant's successful integration into a new group (Strum, 1982). As noted in Chapter 6, lactating females, infants, and juveniles are extremely wary of new males. They generally avoid them, and avoidance is frequently accompanied by raised tails or screams, responses that draw the attention of other troop members, including resident males, to the newcomer (Packer, 1979a; Busse, 1984a). Aggression by resident males, combined with the vigorous negative responses of mothers, make it difficult for newcomers to pursue normal activities, such as feeding and resting, near other troop members. This reinforces their peripheral and unfamiliar status.

Newcomers sometimes try to overcome this problem by forming consortships with females during the early tumescent phase of the estrous cycle (Packer, 1979a). These females, unlike mothers, often show active interest in newcomers and may cooperate in maintaining proximity to the males. Packer found that aggression by resident males decreased when newcomers were consorting, and it was my impression that mothers were less wary of consorting newcomers, although quantitative data on this point are lacking.

Packer (1979a) reported that newcomers abandoned such consort partners once the resident males moved away, but in EC, a loose association between a newcomer and an estrous female often persisted for several days, or even weeks. These associations resembled those between a resident male and a noncycling female who were Friends, except that the newcomer seemed to take more responsibility for maintaining proximity than resident males normally do (see Chapter 5). These relationships between newcomers and cycling females appeared instrumental in allowing the newcomer to maintain relaxed proximity to other members of the troop, an achievement that probably contributes to his acceptance as a new troop member by both males and mothers. Males who were unable to form relationships with particular females seemed to have a much more difficult time becoming integrated into the new troop.

Could female acceptance also influence how long males remain in a troop? There is evidence that it does. Table 9.1 shows how often anestrous females exhibited long-distance avoids in response to approaches by males. As noted in Chapter 6, females avoided newcomers most often, and then short-term residents, and they avoided long-term residents the least. However, within each age/residence class, there

Table 9.1. Female Responses to Males and Length of Male Troop Tenure[a]

Male age/residence status at start of study	Male	Percentage of male "approaches" that elicited a female long-distance avoid	Date male disappeared from troop[b]
Newcomer	AA	16.0 (75)[c]	Still present
	TN	29.7 (37)	December 1980
Short-term resident	SK	3.5 (85)	Still present
	AC	10.8 (93)	November 1979
	HD	14.7 (95)	November 1979
	IA	23.8 (21)[d]	August 1978
Long-term resident	BZ	1.1 (188)	Still present
	HC	3.7 (107)	Still present
	AG	6.4 (47)[d]	August 1978
	CY	7.6 (131)	July 1979
	VR	8.9 (90)	August 1979

[a]Values are based on 790 hours of focal samples on pregnant and lactating females. "Approaches" were movements by the male that brought him to within 5–15 m of the female. If she immediately moved away from him, a long-distance avoid was scored.

[b]Still present indicates that the male was still a member of EC in May 1983.

[c]Number in parentheses indicates total number of "approaches" by the male.

[d]These males have far fewer approaches than other males of the same age/residence status because they disappeared before the study was completed.

were striking differences in how often females avoided males, and for all three classes, the males that females avoided least in 1977–1978 were the males who subsequently remained in the troop the longest.[2]

Descriptions of the careers of two pairs of males who transferred into EC illustrate these findings. Ian and Handel transferred to EC from their natal PHG in April 1977 (S. Strum, personal communication). By the time my study began 5 months later, Handel had begun to establish friendships with two females. Over the next few months, his consort activity steadily increased (from September 1977 through August 1978, he ranked fifth among males in consort success during conception cycles), and he added three other females to his list of close associates.

Ian, on the other hand, had great difficulty establishing relation-

[2]The circumstances in which the other males disappeared were not known. It is possible that VR was poisoned by farmers, but we suspect that all other disappearances were voluntary departures, since the males showed no signs of ill-health prior to departure.

ships with females. His tentative approaches to females almost invariably provoked alarm: The females would geck, fear-grin, raise their tails, and prepare to flee. Most males of Ian's age (around 10 years) have already learned how to appease frightened females. They cease approaching, sit down, and reassure the female with friendly gestures and sounds, such as the "come-hither" face, lip-smacking, and soft grunts. This does not always work, but often the female will calm down and allow the male to approach; sometimes she will even approach him. Ian, however, seemed to lack entirely the social skills necessary to placate nervous females. When a female showed alarm, he would continue his approach, and when she turned to flee, he would chase her. Sometimes the female would scream, and this tended to evoke a mobbing response from nearby animals. I saw Ian chased out of the troop by a horde of females and juveniles on several occasions. As a result, he spent most of his time on the edge of the troop, far from other baboons. His relationships with EC males did not cause Ian's peripheral status; he was larger than Handel and more intimidating, but his extreme lack of popularity with females kept him from becoming fully integrated into the troop. He disappeared in August 1978 and probably transferred to another group.

Alex and Triton present a similar contrast. Both entered EC in July 1977 as subadult males (S. Strum, personal communication). Both were unusually large males by March 1978 (see Table 2.4), and each was capable of holding his own against most of the EC males. Alex had a remarkably relaxed disposition and rarely evoked alarm in EC females, but he only slowly began to form friendships, perhaps because he was very passive and seldom initiated interactions with anyone. However, after 1 year in the troop, females began to initiate friendships with him, and when I returned to EC in July 1979, he had six female Friends and was forming consortships with a variety of females. Alex was still a successful consorter in July 1983, 6 years after he first entered EC (Smuts and Watanabe, in preparation).

Triton, in contrast, frequently aroused fear in EC females, although he was less peripheral than Ian. He tried to form friendships with females, but he was not very successful, and by July 1979 he had only one female Friend. His consort activity remained low, and he disappeared from EC in December 1980.

Why some young newcomers are so much more successful than others at cultivating bonds with females is unclear. When observing different males interacting with females, one cannot escape the impression that individual personality traits, such as confidence and sensitivity to the female's response, are the most important factors determining female acceptance. Developing consistent measures of relevant personality differences presents an important challenge for future research (cf. Stevenson-Hinde, 1983).

How New Males Make Friends

Below, the development of friendships by two short-term residents, Sherlock and Handel, who eventually achieved full integration into EC as well as moderate to high consort success is described, in order to illustrate the process of male integration into a new troop. It is based on qualitative observations, and my impressions need to be confirmed by more quantitative data.

Handel and Sherlock appeared to take a more active role in forming friendships with females than did older, long-term resident males. While initiating a relationship with a female, these short-term residents were more responsible for maintaining proximity than was the female—a reversal of the pattern in established relationships where the female accounts for more approaches and fewer leaves than does the male (Chapter 5). Handel and Sherlock also groomed females more while initiating relationships than they did once the relationship was established. After this period of active "pursuit" of particular females, lasting anywhere from several weeks to a few months, the male and female roles in maintaining proximity sometimes changed, and the female began to follow Handel or Sherlock. This transition marked the establishment of a long-term bond (S. Strum, personal communication). If this shift did not occur, it seemed to indicate the female's lack of "acceptance" of the male, and he would shift his attentions to another female.

At first, both Handel and Sherlock focused their attentions on young (although not adolescent) females. Only after each had developed stable relationships with such females did he attempt to cultivate relationships with older ones. At first, neither male interacted much with lactating females who had young infants. Instead, they developed relationships with cycling females, and they tended to associate with these females primarily during the early phases of sexual swelling and nonswollen phases of the cycle when other males were not forming consortships with the females, thereby avoiding intense competition with other males. After Sherlock and Handel had succeeded in forming consortships with females during peak estrus, they began to associate with mothers who had young infants, usually the same females they had consorted with in the past. As mentioned in Chapter 6, females with young infants are wary of males who were not in the troop at the time the female conceived (Packer, 1979a; Busse and Hamilton, 1981; Busse, 1984a). My observations on the female associates of Handel and

Sherlock suggest that female wariness sometimes also extends to males who were in the troop when the females conceived, especially if these males were recent immigrants at the time and if they had not consorted with the females in the past.

It would be interesting to compare the development of friendships involving males like Sherlock and Handel with those involving long-term residents. Unfortunately, most of the friendships involving these older males were already established when my study began, and the process by which the others were formed was not obvious. I suspect that I failed to discern the early stages of these relationships precisely because they were based on more subtle, reciprocal behavioral interactions than the new friendships of recently transferred males.

THE IMPACT OF SEX: MAKING AND LOSING FRIENDS

After a baboon female conceives, her estrous cycles cease, and, unless her infant dies, she will not be sexually receptive again for 1–2 years. In EC, this interval of complete sexual abstinence lasted, on average, 20 months (Nicolson, 1982). During this period, friendships with males remained remarkably stable. Eight females who were lactating when the first phase of my study ended (December 1978) were still lactating and had not resumed estrous cycles when I returned to EC, 7 months later. For all eight, the favorite grooming partner after my return was one of the female's Friends during the previous study period.

Throughout this period, friendships are entirely platonic: Sex plays no role in maintaining the bond. At the same time, neither sexual attraction to other males nor sexual interest from other males threatens the relationship. Then one day, estrous cycles resume, and, literally overnight, the chaste mother is transformed into an alluring female with a strong sexual appetite. Suddenly she begins to interact with a wide variety of willing males. What happens to her old friendships? How disruptive is sex? What role does it play in the creation of new friendships and in the demise of old ones?

Although there has been no systematic research on these questions, the first set of observations described below suggests that answers to these questions vary, depending on the female, her Friend, and her consort partners. The second set of observations focuses on how new bonds are formed during periods of estrous cycling.

What Happens to Friendships When the Female Resumes Estrous Cycles?

Eight EC females were observed during three consecutive reproductive phases: (1) pregnancy and lactation; (2) subsequent estrous cycles;

and (3) the following period of pregnancy and lactation. Table 9.2 shows these females, their Friends during both phases of pregnancy and lactation, and their consort partners during the intervening period of estrus.

Three main findings emerge. First, most anestrous friendships (12 of 13 in which the male remained in the troop) persisted into the next period of pregnancy and lactation. Second, prior bonds were likely to persist through cycling and into pregnancy and lactation even when the female consorted with other males, but not her Friend, during the conception cycle (5 cases). These results are consistent with Ransom's characterization of "pair bonds" among Gombe baboons: "This bond, an intense, long-lasting relationship between an adult male and female, appeared to be unaffected by certain temporary aspects of the participants' behavior, such as the female's estrous cycle and the male's consort activity with other females" (1981, p. 233).

The third finding, however, suggests that there are some exceptions to this generalization. After conception, three females showed evidence of a new bond with a male they had consorted with, and one of these bonds (EU and HD) appeared to "replace" a previous friendship (EU and HC), perhaps in part as a result of HC's failure to consort with EU. It is important to note, however, that in the majority of cases (8 out of 11), consort partners who were not already Friends did not become Friends during subsequent pregnancy and lactation. These results indicate that sexual interaction sometimes, but by no means always, leads to the formation of new friendships.

It is possible that the EC data underestimate the potentially disruptive effects of estrous cycles on long-term bonds for three reasons. First, the data were collected during a 2-year period when no males transferred into the troop. This is atypical: Several males transferred into EC in the years preceding and following the study (S. Strum and J. Johnson, personal communication). During times when male membership is less stable, friendships may be more vulnerable to disruption by estrous females' attraction to new males (cf. Packer, 1979a). Second, the results are based on only eight females, five of whom were classified as "older middle-aged" females (see Table 2.3). I had the impression that friendships of older females were generally more stable than those of younger females. If this is the case, then this small sample may be biased toward more stable relationships. Third, if, as my results indicate, new friendships sometimes develop as a result of consort activity, then either: (1) Females accumulate more and more friendships as they age, or (2) some friendships must end. Since older EC females did not have more Friends than younger females (Chapter 4), some friendships must end.

Table 9.2. Comparison of Friendships of Anestrous Females Before and After Intervening Periods of Estrus and Consort Partners during Estrus[a]

Female/male dyad		Friends Phase 1: Pregnancy and lactation	Consort partners Phase 2: Estrous cycles	Friends Phase 3: Pregnancy and lactation
AM	AO	+	+	+
CI	HC	+	+	+
EU	VR	+	+	+
HH	BZ	+	+	+
HH	SK	+	+	+
PO	SK	+	+	+
ZD	HC	+	+	+
AM	BZ	+	−	+
AT	HC	+	−	+
CI	VR	+	−	+
JU	SK	+	−	+
ZD	AO	+	−	+
EU	HC	+	−	−
AT	BZ	−	+	−
AT	SK	−	+	−
CI	HD	−	+	−
CI	TN	−	+	−
HH	CY	−	+	−
JU	VR	−	+	−
PO	BZ	−	+	−
ZD	CY	−	+	−
EU	HD	−	+	+
JU	HD	−	+	+
PO	HC	−	+	+
PO	HM	−	−	+
AT[b]	CY	−	+	0
PO[b]	CY	+	−	0
AT[b]	AG	+	0	0
EU[b]	AG	+	0	0

[a]Phases 1, 2, and 3 follow each other in time. Phase 1 includes pregnancy for all eight females and lactation for six (one female miscarried and one had a stillborn infant). Phase 3 includes pregnancy for all females and the first few months of lactation for six. Plus = the relationship existed; minus = the relationship was absent; zero = the male disappeared from the troop. For seven females, consort partners included all males observed consorting with the female from D-7 through D-1 of her conception cycle. The conception cycle of the eighth female, EU, was not observed, and data are from the previous two cycles. Identification of Friends for Phase 1 was based on methods described in Chapter 4. For Phase 3, friendships were based on ad lib grooming records only.

[b]These four dyads involved males who disappeared after Phase 1 or Phase 2, as indicated by a zero in the appropriate column.

Comparable data on the persistence of affiliative relationships through the three reproductive phases listed above are available from only one other study, for two females (Seyfarth, 1978a,b). One of these females maintained her old bond even though she consorted with a different male. The other did not consort with her previous affiliate, and, after conception, her old bond was replaced by a relationship with her consort partner. These results show that, in other troops as well as in EC, the effect of estrous cycles on previous bonds varies.

Courtship of Estrous Females

All studies of baboon sex have focused on consortships and copulations during peak estrus. This is when male-male competition over females is most intense, and it is also the period when ovulation is most likely to occur (see Chapter 7). From a strictly reproductive point of view, this is the time that matters, and the vast majority of consortships and copulations between adult males and females occur during this interval (Scott, 1984). Rapid detumescence of the sexual swelling marks the end of this period—a signal that the female has passed the ovulatory stage (Hendrickx and Kraemer, 1969). Typically, the female's interactions with males undergo a sudden shift around this time. The males who were vying intensely for opportunities to mate with the female only a day or two earlier now show little interest in her, and the female, after being carefully guarded for several days, is suddenly free to copulate with any male. While her swelling is shrinking, the female continues to show an appetite for sex, but most of her copulations are with young males, who are denied access to her during peak estrus by the possessive behavior of the female's consort partners.

There are, however, important exceptions to this general scenario. Although copulations almost always cease shortly after D-day (the first day of rapid detumescence of the sexual swelling), sometimes an adult male consort will continue to follow the female, groom her frequently, and behave possessively when other males are nearby. In some cases, these "pseudo-consortships" persist right through the stages of detumescence, menstruation, and inflation of the sexual skin, until the female again reaches peak estrus. At that point, the dogged suitor must once again compete with other males for access to the female, but such persistent courtship appears to increase a male's chances of maintaining a consort relationship during the next cycle (Collins, 1981).

During a pilot study of olive baboons in Masai Mara Game Reserve, I was struck by the frequency of pseudo-consortships, and for a brief period I was able to collect quantitative data on these relationships. Focal samples were conducted on three cycling females well outside the

Table 9.3. Comparison of Spatial Proximity between Males and Cycling Females When the Females Were in Peak Estrus ("Consort Pairs") and at Other Times ("Other Pairs")[a]

Behavioral measure	Type of pair: Consort pairs	Other pairs
Percentage of time spent at different distance categories[b]		
Within arm's reach (1 m)	42.9	42.0
Between 1–5 m	37.5	38.4
Greater than 5 m	19.6	19.6
Percentage of time the male of the pair was the female's nearest neighbor	89.3	84.5
Male responsibility for maintaining proximity (% approaches by the male minus % leaves by the male)	79.2	62.9
Percentage of minutes of grooming between male and female where male groomed female	53.2	88.1

[a]Values are based on 21 10-minute focal samples of three consort pairs and 13 10-minute samples of three different "other" pairs involving females outside of peak estrus. This period covered about 15 days and included three phases: late detumescence, a few days before menstruation; the "flat" phase of the cycle, when menstruation occurs; and early tumescence, the first few days of sexual swelling.

[b]Proximity was recorded every minute, on-the-minute, throughout the 10-minute focal sample.

period of peak estrus. For comparative purposes, I conducted similar samples on females during peak estrus who were involved in true sexual consortships. (Due to logistical difficulties, it was not possible to collect data on the same females throughout the estrous cycle.)

Table 9.3 compares the pseudo-consort pairs and the true consort pairs on several behavioral measures. The two types of pairs did not differ in the amount of time the male and female spent at the three distance categories shown in the table, and in both, the male was the female's nearest neighbor most of the time. Males were primarily responsible for maintaining proximity in both types of pairs, although they took the initiative slightly more often during true consortships. In both types of pairs, males groomed females considerably more than males normally groom females (see Chapter 5), and this tendency was even stronger for males in pseudo-consort than for males in true consort relationships. In fact, aside from the stage of the female cycle in which they occurred, pseudo-consortships were distinguishable from true consortships in only two ways: The animals did not copulate, and other males did not harass the "consorting" pair so often.

Despite reduced harassment, however, males in pseudo-consortships do show possessive behavior toward the female, just as males in true consortship do (Collins, 1981; Smuts, personal observation). These behaviors include herding the female away from other males and demonstration of "possession" through intense grooming and mounting of the female (without intromission) when other males are nearby.

Andromeda and Dante, a young, high-ranking male, are feeding together. During Andromeda's period of peak estrus, which ended about 1 week earlier, Dante and four older, long-term resident males competed intensely for access to her (see examples in Chapter 7). Ever since D-day, Dante has been associating with Andromeda. On this occasion, the baboons are feeding peacefully, and Andromeda begins to wander away from Dante. He glances at her every few minutes, keeping careful track of her whereabouts. Some 20 minutes pass, and Andromeda is now about 50 m away. Dante continues to monitor her movements, but he does not seem concerned that she is so far away. Then Boz, one of the older residents who competed for Andromeda earlier in her cycle, approaches to within about 15 m of Andromeda. Dante immediately runs over and interposes himself between Andromeda and Boz. He grooms her intently for a few minutes, glancing frequently at Boz, who ignores the grooming pair. Andromeda moves away to continue feeding, but Dante continues to watch her carefully for several more minutes. Then Boz moves away, and Dante, too, resumes feeding (focal sample on Dante, 21 June 1983).

Later that same day: Andromeda again wanders away from Dante. This time she is obscured from his view by a small hill. Andromeda is attacked by a higher-ranking female, and she screams loudly. Dante, apparently recognizing her voice, runs about 50 m to the scene and chases the other female away. After the chase, Andromeda approaches Dante and grooms him for 10 minutes (ad lib observation, 21 June 1983).

These examples illustrate the methods males use to form bonds with estrous females. They combine behaviors used to establish and maintain true consort relationships (e.g., preventing the female from interacting with other males) with behaviors they typically show toward Friends (e.g., protecting the female from aggression by other baboons). The possessive behaviors signal the male's high motivation to maintain an association with the female, and they inhibit rivals' attempts to interact with the female both during and outside of peak estrus (Collins, 1981). The friendly behaviors, such as protection and grooming, are probably important because they show the female that this male is both willing and able to provide the benefits associated with long-term bonds. If these demonstrations increase the female's preference for the male as a mate, they can increase his chances of consorting with her in two ways. First, she may become a more cooperative consort partner (Rasmussen, 1980; Collins, 1981). Second, her preference for him may reinforce rival inhibition, since rivals are least likely to challenge possession of females who prefer their current partners (Bachmann and Kummer, 1980; see Chapter 8 for further discussion).

Evidence from EC supports the hypothesis that persistent association throughout all phases of the estrous cycle increases a male's chances of consorting with a female during peak estrus. Three females, PS, ZN, and TH, cycled for most of the study, conceiving near the end. Data on grooming and on proximity to males when they were cycling but not in consortship were used to determine the primary male associates of these three females. Among them, the females had five associates, and in every case, the associate consorted with the female more often than expected based on his overall consort activity [*1*]. Thus, in the short-term, persistent courtship of females throughout all phases of the estrous cycle appears to be a tactic males use to increase opportunities for mating at peak estrus.

In some cases, however, pseudo-consortships have long-term effects as well. Most of the new friendships that develop during estrus are probably a result of such persistent associations. A good example is the relationship between Justine and Handel (Table 9.2). Prior to the resumption of cycling, Justine rarely associated with Handel (he ranked 11 out of 18 males on the composite proximity score described in Chapter 4). During her first two estrous cycles, HD was the only adult male who showed much interest in Justine. He formed long consortships with her, and they continued to associate between her periods of maximal swelling. During Justine's conception cycle, several adult males competed for access to her, and she consorted with at least one male besides Handel. However, Handel spent more time in consort with Justine than any other male, and he was the most likely father of her next infant. I returned to EC shortly after her infant was born, and it was evident that she and Handel had formed a new friendship.

HOW LONG DO FRIENDSHIPS LAST?

My return to EC in 1983 gave me an opportunity to find out whether any of the friendships I had identified in 1977–1978 were still intact. Unfortunately, only three of the original adult males were still in EC (Alexander, Boz, and Sherlock). Among them, they had had 16 Friends, 10 of whom were still alive in 1983. Based on ad lib grooming records and spatial proximity during focal samples on males, it was evident that at least five of these friendships had survived 6 years: three involving Boz and one each for Alexander and Sherlock. In two other cases, Boz's old friendships were no longer apparent, but these females were cycling, and it is sometimes difficult to determine the long-term bonds of females in estrus. In three other cases, all involving Sherlock, the friendships appeared to have ended. However, even in these cases, I did not feel certain of my conclusions because of limited time and difficult observation conditions.

The important point, however, is that each male retained at least one friendship from the earlier period. It is not known how long baboons live in the wild, but 25–30 years is a reasonable estimate for females (males may not live so long). Based on this estimate, 6 years represents from one-quarter to one-third of the life of an adult female baboon—a long time for a relationship to last. And this, of course, is a minimum estimate. It is possible that friendships occasionally last for life, but it will require longer studies and some very patient scientists to find out.

EMOTIONS UNDERLYING SEX AND FRIENDSHIP

Introduction

In the second half of this book, the reproductive benefits that females and males derive from friendships with one another have been discussed in some detail. Investigation of these benefits clarifies the evolutionary basis of friendship-forming tendencies in baboons, but it does not tell us what motivates individual baboons to make friends. Baboons do not decide among alternative courses of action by evaluating their contributions to reproductive success, and neither do we. To understand the immediate causes of human and animal behavior, it is necessary to shift levels of analysis. We need to consider motivations: changes in internal states that direct the individual toward some goal that will decrease negative feelings (e.g., hunger, fear, loneliness) and/or increase positive ones (e.g., satiation, security, sexual pleasure).

Motivations and the subjective, emotional states that accompany them are mechanisms that have evolved through natural selection to make individuals behave in ways that, on average, increase reproductive success: Hunger motivates animals to find food, sexual desire leads to copulation, and attachment to young results in caretaking behaviors that promote offspring survival (Hamburg, 1968). What, then, are the motivational mechanisms that lead baboons to make and keep friends?

Although this is an important question, it is a difficult one to answer. In all creatures, including humans, motivations and emotions can only be inferred from behavior. When dealing with people, of course, we can use language to gain insight into another's subjective states, but even then, we cannot have any direct experience of what someone else feels. With animals, the problems associated with inferring emotion are magnified, both because we cannot use language to converse about internal states, and because we are not members of the same species; the latter is a handicap because the less familiar we are with another being, the less skillful we are at inferring emotions from behavior.

Despite these difficulties, people infer motivations and emotions in animals all the time, and often with great success, to judge by the results. People who live around wild animals learn to distinguish between hungry predators, which are dangerous, and satiated ones, which are not. People who work with domestic animals learn to respond to subtle cues that indicate an animal is angry and likely to attack or frightened and about to flee. And, of course, pet-owners derive great rewards from their ability to discern a wide variety of internal states from behavioral cues.

Using similar skills, scientists routinely infer animal motivational states like hunger and fear. However, it is usually not considered appropriate to extend this process to include internal states associated with social interactions, such as affection or ambivalence, trust or jealousy. Scientists are generally most concerned about the dangers of anthropomorphism in social contexts.

This concern is sometimes valid, but too often it has led students in animal behavior to shy away from intriguing and important questions (Griffin, 1984; de Waal, 1986). When we refuse to use our inferential skills to investigate animal emotions, we forgo knowledge of the forces that give animals personality and vitality; we deny them the spirit that has captured our interest to begin with. This seems a large price to pay for the certainty and safety of "objective" knowledge.

In what follows, I give several examples of behaviors associated with baboon friendship that seem best described in terms of familiar human emotions. It is necessary to stress that use of these terms is not meant to imply that the behaviors described are thereby understood and hardly worthy of further scrutiny. My goal is just the opposite: to stimulate consideration of a whole universe of social interaction the existence of which has barely been acknowledged as a legitimate subject of scientific inquiry.

Ambivalence

In EC, prior bonds facilitated consort formation, which suggests that females often preferred to mate with the males who were most familiar to them, the males with whom they were most at ease. Yet, there is strong evidence that estrous females are sexually attracted to strange males (Packer, 1979a; Chris and Aaron example, this chapter). Which males do female baboons really prefer?

I do not think that this question can be answered. Females are drawn to both types of males, and their behavior suggests that they frequently experience conflicting impulses. These conflicting desires are most transparent when a female with a sexual swelling has an opportunity to interact with a strange male. She will begin to approach him, stop, look around her, take a few more steps, and then pause again. If he makes a move toward her, she will probably run away. If he remains still, she is likely to continue her approach. Sometimes, she loses her nerve before she reaches him, and often a long series of unfinished approaches precedes the actual contact.

There is tension and ambivalence in long-term relationships as well. Females seem to feel more secure when they are near their male Friends (see below), but they are also subjected to annoying behaviors. Males constantly appropriate the feeding sites of nearby females. This can be particularly exasperating when the female spends several minutes digging for corms, only to have the food repeatedly appropriated by the male. Once, I saw Phoebe cease feeding after her Friend Cyclops had supplanted her a dozen times in the space of 5 minutes. She watched him for a few moments and then with great purpose, gathered her infant and walked to a new site, far away from her protector. A few minutes later, Phoebe was chased by a higher-ranking female, and after the chase was over, she sat and looked around, as if carefully weighing her alternatives. She scratched herself several times, as baboons often do before making a decision, and then returned to Cyclops.

Conflicting emotions are not restricted to females encountering new males or deciding whether to feed near old Friends—in fact they seem to be a continuous part of every baboon's daily existence. When one considers one's own experience, this is hardly surprising: Ambivalence is a more or less inevitable aspect of social interaction. We are drawn to new people but feel shy or nervous around them. We consider becoming involved in a new relationship but are concerned that it will threaten an old one, and so on. Baboons, like humans, are enmeshed in a web of social relationships from birth until death. Yet, in both species, the same individuals who are an individual's closest companions, mates, and protectors are also sources of competition, fear, and sometimes even death. Natural selection has provided humans, baboons, and other intelligent, social animals a subjective state—ambivalence—that mirrors the ambiguities and uncertainties of the social world around them. It is hard to imagine how it could be otherwise.

Flirtation and Courtship

Very little is known about how baboon friendships develop, and so this section must be brief and tentative. However, on the few occasions when I witnessed a series of interactions that, with hindsight, turned out to represent the early stages of a friendship, I was struck by the parallels with human courtship.

To begin with, baboons flirt. One of the best examples of flirtation is the interchange between Thalia and Alexander described in the Prologue. I have seen several such exchanges, and all of them took place on the cliffs, when the baboons were well fed and relaxed. In each case, eye-contact was the key. First, the male and female looked at one another but avoided being caught looking. This phase was always accompanied by feigned indifference to the other, indicated by concentrated interest in grooming one's own fur or by staring intently at some imaginary object in the distance (both ploys are commonly used by baboons in a wide variety of socially discomfiting situations). Eventually, the coy glances were replaced by a more direct approach, usually initiated by the male. Grooming then followed. If the friendship "took," the couple would then sit together during daily rest periods, and eventually they would coordinate subsistence activities as well. However, I have no idea how many flirtatious exchanges actually result in a long-term bond. Perhaps, as in our own species, most do not.

In other cases, indications of attraction were more direct. Chris's behavior when luring Aaron into her troop (this Chapter) is a good example. Sometimes, a friendship seemed to develop as the result of persistent, almost obsessive courtship by the male; the female was not necessarily cycling at the time. The male's attentions intensified quite suddenly, as if he had become the victim of a "crush." He followed the female about, interacting with her whenever she was willing. When he was prevented from interacting, either because she was with another male or because she resisted, he would sometimes just sit and watch her. Did he lose his appetite? Probably not, but we will never know unless we begin to look for subtle indications of a change in emotional state in individuals who are forming, or ending, relationships.

Jealousy

The baboon literature is replete with examples of males behaving possessively toward their consort partners, and Collins (1981) reported that males show possessive behavior toward estrous females at all phases of the cycle (see earlier section on "Courtship of Estrous Females"). During focal samples on pregnant and lactating females, I saw male Friends behave in an obviously possessive manner only 12 times, or once about every 60 hours (possessive behavior was scored, for example, when the male chased his Friend away from another male or ran over to groom her right after she presented to another male). I suspect, however, that these observations grossly underestimate the frequency of possessive behavior toward anestrous females for two reasons. First, many possessive behaviors by long-term resident males were elicited by recently transferred males' attentions to females, and these same males were also more likely than older males to behave possessively toward females. Because no new males transferred into EC during my study, possessive behavior was probably less frequent than it is at other times. Second, unless the observer is focusing on possessive behaviors, she or he will probably miss all but the most obvious interactions. This fact was brought home to me with force on a rare occasion, described below, when it was possible for two observers to watch the male and female members of a friendly couple at the same time.

A fellow primatologist, Richard Wrangham, was visiting EC. A few baboons had begun to ascend the sleeping cliffs, and we went with them. We settled 2 m from Sherlock, a short-term resident male, who sat, stomach full, holding his toes, looking sleepily out over the field below, where the remainder of the troop was still foraging. Cybelle, a lactating female with whom Sherlock was developing a new friendship, came into view, and I noticed that he suddenly became more alert. I suggested to Richard that he keep an eye on Sherlock, while I watched Cybelle, to see if Sherlock responded in any way to her movements or behavior. For a few moments, Cybelle sat quietly, nursing her baby, and Sherlock's lids began to droop. Then a female approached Cybelle and pulled on her infant, and the baby lost the nipple. Sherlock, much too far away for Cybelle to hear him, grunted as the infant lost its hold; this was not too surprising, since baboon vocalizations often reflect a vicarious involvement in the interactions of others. Cybelle got up to avoid the female who was handling her infant, and as she moved, she walked past an adult male, Hector, who although not a Friend, had shown some interest in Cybelle after her baby was born. At exactly this moment, Richard, eyes on Sherlock, said to me, "What is she doing? Sherlock's body has gone all tense and he's staring at something." Cybelle continued walking past Hector, and as she did so, Richard said, "Something's changed; he's relaxed again." A few minutes later, Cybelle actually presented to another male. Again, Richard had no idea what Cybelle was up to, because he was watching Sherlock. "He's leaned forward," he said, "and now he's standing up." Sherlock began to pant-grunt (a vocalization used in threat) under his breath, and, according to Richard, looked as if he were about to rush headlong down the cliff. Just then, however, Cybelle moved away from the other male and glanced up at Sherlock. "He's making a very funny face, indeed," said Richard, describing to me the most intense version of the "come-hither" face: ears back against the skull, eyes narrowed, chin and shoulders pulled in, and lips smacking madly. Cybelle responded with a much milder "come-hither" face and then resumed foraging. I said to Richard, who had begun to empathize with Sherlock's anxiety over Cybelle, "Don't worry. That means she'll probably be up here soon." My predictions are not always correct, but in this case, I was right, and Richard was duly impressed. Cybelle fed for about 5 minutes more and then joined Sherlock on the cliffs. They took turns grooming each other until it was time for us to leave (ad lib observation, 23 December 1977).

Females also show evidence of jealousy. Seyfarth (1976) found that 25% of all agonistic interactions between females were over access to males as social (not sexual) partners. Competition for males was not so obvious in EC: Less than 5% of female agonism occurred in this context. This difference probably reflects the sex ratios of the two groups. In Seyfarth's group, there was one adult male for every four females, while in EC, there was one adult (or large subadult) male for every two females. Nevertheless, I saw females threaten, chase, and even attack other females who were interacting with the aggressive female's Friend.

Trust

One day when I was looking for EC, I came across Louise and her Friend Virgil, all by themselves, feeding in an open field. Louise was still lactating, but her infant was more than 1 year old and had remained with the rest of the troop. Finding Louise and Virgil alone together was a surprise, because, except for males transferring between groups and occasional consort pairs, baboons very rarely travel apart from their troop. At first, I thought that perhaps these two had somehow become accidentally separated from the others, but they spent the whole day alone together and made no attempt to find their troopmates, who, it transpired, were about 2 km away. As night drew near, Virgil and Louise joined the rest of the troop at the sleeping cliffs.

Although predators large enough to capture an adult female baboon are rare near Gilgil, the troop fled from people and dogs as if they were predators, and, in general, EC baboons seemed no less alert to possibilities of danger than baboons I had watched in areas populated by leopards and lions. I remember thinking at the time that Louise must have felt very safe in Virgil's company in order to risk a day-long separation from the rest of the troop; she must have trusted his ability and willingness to protect her from danger.

On several other occasions, I saw a male leave the group to forage on his own for a time, and in each case, one or two females followed him. They were always his Friends. With one unusual exception, however, I never saw a female away from the troop on her own without a male Friend.[3]

Female trust in male Friends was also apparent in their willingness to leave young infants with the males while they foraged, sometimes several hundred meters away. Ransom (1981) refers to this habit as male "baby-sitting." The Friends kept a careful eye on the infants and made sure that they were not harassed by other baboons. Usually, after an hour or so, the mother would return, or the male would lead the infant back to the mother, but on a few occasions, the troop fled from a person or predator before the mother and infant were reunited. In each case, the male rushed to the infant and carried it until the alarm had passed. The only other baboons I have seen perform the "baby-sit-

[3]For several years, a very old EC female, Athena, disappeared from EC for weeks at a time. Usually, we had no idea where she went, but twice we found her alone, and another time she spent a day with the adjacent troop, PHG (D. Manzolillo, personal communication). Athena behaved oddly when with EC, and we think she probably suffered from some form of senility. She disappeared for good in 1980.

ting" role, including prolonged carrying of an infant separated from its mother, were siblings of the infant. This suggests that, when it comes to infant welfare, females consider their male Friends as trustworthy as a close relative.

Affection

By our standards, baboons are relatively undemonstrative toward their Friends. They rarely hug one another, and, even after being apart for several hours, physical expressions of attachment are uncommon. This is not because hugs and kisses are not part of their repertoire: Adult females routinely hug other females and juveniles in greeting, and females occasionally kiss by bringing the nose or mouth against another's face. Does the rarity of such gestures among Friends reflect a less affectionate relationship?

Once again, a comparison of Friends and relatives is useful. Close kin—mothers and offspring or siblings—almost never greet one another. In fact, their interactions are remarkably devoid of the gestures that unrelated animals routinely use to communicate intentions and attitudes during social interaction (Smuts, in preparation). This suggests to me that many baboon gestures, including greetings, are equivalent to human conventional courtesies; their function is to smooth the way for social interactions between individuals whose relationships remain ambiguous. In baboons, as in humans, members of the immediate family and close friends have less need of such gestures. In the example that follows, interactions with several nonfamily members are compared with those involving family or Friends.

Cybelle is grooming her juvenile female daughter, Minthe, while her older daughter, Delphi, and her subadult son, Socrates, groom next to them. Cybelle's 3-month-old infant, Cera, clings to her tummy. A lower-ranking adult female, Lysistrata, approaches. After watching the family for a moment, she presents her lowered rump to the mother and infant and lip-smacks and grunts at Cera. While in this position, she steals several quick glances at Cybelle, as if to make sure her approach is acceptable. Cybelle at first ignores her, but when Lysistrata reaches to touch Cera, Cybelle raises her brows slightly, and this mild threat is enough to send Lysistrata away.

A few moments later, another adult female, Helen, approaches. Because she ranks higher than Cybelle, she does not begin the interaction by presenting her lowered rump. She instead walks straight up to the pair, and, looking alternately at

Cybelle and Cera, she grunts and lip-smacks. These vocalizations indicate to Cybelle that Helen wants to interact with Cera, and they reassure Cybelle that she does not need to move away in order to show deference to Helen. Nevertheless, all members of Cybelle's family stop grooming while Helen is near, and only after she leaves do they resume their former activities.

Helen's approach is followed by that of Virgil, an older resident male who is not a Friend of Cybelle's. He pauses as he walks past and grunts at the infant. Cybelle and the others again stop what they are doing, and Cybelle gathers Cera close to her and prepares to leave. Virgil moves on, and the family members resume grooming.

The next animal to approach is Minerva, Cybelle's adolescent daughter. She approaches directly and does not vocalize at all. First, she tugs on Cera. Cybelle pushes her hand away, but she continues to play with the infant. Then she pushes in between Cybelle and Minthe, who is being groomed by Cybelle, and presents her own flank for grooming. Cybelle is forced to stop grooming Minthe, but she ignores Minerva, and Minerva then walks over to her older sister, Delphi, and her brother Socrates, who are also grooming. She squeezes in between them and then pushes Delphi to the ground, holding her there with one arm while she presents to Socrates for grooming. Socrates grooms her, and she releases Delphi, who has remained passive throughout. Delphi shifts her position and begins to groom her mother.

The tranquil scene is completed when Sherlock, one of Cybelle's Friends, approaches. Like Minerva, but unlike the others who approached, he moves directly toward the family group without pausing and plops himself down in front of Cybelle. She grooms him while Cera climbs on his back. The others ignore Sherlock, continuing their previous activities (focal sample on Cybelle, 8 November 1978).

This example illustrates an important principle: Among baboons, the clearest indication of a strong, affectionate relationship is routine association and relaxed demeanor, not physical expressions of affection. Baboon Friends, however, occasionally do express affection overtly, not in the context of greetings or other "conventional" interactions but spontaneously, as if from the heart.

Virgil, an older, resident male, is sitting and resting. He is the center of a play-group of a dozen juveniles and infants. A few feet away, Isadora, one of Virgil's Friends, is grooming with several members of her family, including her daughter, Venus. Venus has just had her first sexual cycle, and, after 3 weeks of playful sex with immature males, her swelling has receded. During this period, she did not interact much with Virgil, whom she has probably known well since infancy. After watching Virgil and the play-group for a few moments, Venus gets up, approaches Virgil from behind, and attempts to encircle his large, furry body in her arms. Virgil, startled, jerks around to see who has dared to disturb his nap. When he sees Venus, he nuzzles her neck, and she grooms him (focal sample on Isadora, 25 November 1977).

The troop has been separated into two groups since I arrived at the cliffs early that morning. The animals I am with seem restless, and they frequently look east, the direction they apparently expect the others to be. Iolanthe, a young adult female carrying her first infant, stops feeding and sits, gazing eastward. A few dark spots appear on the horizon; with my binoculars, I can see that they are baboons, but I have no idea which ones. Iolanthe begins to grunt, excitedly, and the others look up. They break into a long chorus of intense grunts, glancing back and forth at each other and the small figures advancing from the east. Suddenly, Iolanthe's grunts become much more intense, and she rushes a few meters away from the group and toward the baboons on the horizon. Looking through my binoculars, I can see that another baboon has come into sight, and I keep my eyes on him for several minutes until I can identify him. It is Boz, Iolanthe's Friend; she apparently had no trouble recognizing him from this great distance. When he and the others are about 100 m away, Iolanthe breaks from the rest of the group and runs to meet her Friend. As she nears him, he slows down and reaches out to touch the infant riding on her back. Iolanthe turns around and falls into step beside Boz. Together, they rejoin the rest of the group (focal sample on Iolanthe, 18 December 1978).

Grief

Do baboons show evidence of grief in response to the death or disappearance of a Friend? I have not detected any such responses, but this does not mean that baboons do not feel sad when they lose someone close to them. Observations of a mother whose infant died illustrate the difficulties one has discerning emotional reactions to loss in wild baboons.

> *Zandra's 3-month-old infant, Zephyr, was killed by a puncture wound in the skull, probably inflicted by a male baboon (see Chapter 6). Zandra carried the body with her for several hours, but then she dropped it in some thick bush and was unable to find it again.[4] She searched desperately for the body for several minutes, but the group moved on, and she eventually followed. For the next several days, I looked for signs of grief in Zandra but could see none. During this time, the troop did not travel near the spot where her infant had been killed and where she had lost the body. Then one day, about a week after her infant died, EC passed through the same bushy area. As they drew near, Zandra became extremely agitated. She rushed about as if looking for Zephyr, and then she climbed a tree. When she got to the top, she looked all around her and began calling. The call she used was the "wahoo" bark, a vocalization that baboons use when alarmed by a predator or when separated from other troop members. This wahoo sounded different, however, from all the others I had heard. It seemed drawn out, plaintive. Perhaps it only sounded different because I suspected that Zandra was thinking of her infant, or perhaps it really was a different sound, and I was able to hear the difference because I was attuned to Zandra's emotional state. In any event, her searching activities were unmistakable, and she repeated her agitated looking and calling each time the baboons passed through this area for the next few weeks (several ad lib observations on Zandra, October and November 1977).*

If I had not been present when Zephyr died and when Zandra lost the body, I would not have understood her strange behavior when the troop returned to the scene, and I would have been unable to detect evidence of an emotional response to the loss of her infant. Cheney (1977b) provides another example of a dramatic response to the death of an infant—in this case by an adult male. An 8-month-old infant chacma baboon was severely neglected by his mother after she resumed cycling. Pierre, a friend of the mother, formed a very close, protective relationship with this infant. The infant died a few months later, and its body lay at the base of the troop's sleeping cliff. Pierre remained

[4]Baboon mothers often carry the bodies of dead infants for several days (e.g., Ransom, 1981; Stein, 1981).

with the body, hardly eating, for two full days while the rest of the troop foraged. The mother showed no interest in her dead infant, but Pierre's closest female associate, Wellesley, helped Pierre to keep his vigil when the rest of the group moved off.

We cannot be sure that baboons mourn the loss of their Friends, but I think it would be premature to claim that they do not.

CONCLUSION

The observations discussed in this chapter lead to four main conclusions. First, relationships between males and females transcend the narrow context of the sexual act. Flirtation, courtship, possessiveness, and jealousy are apparent in interactions between the sexes throughout all phases of the female reproductive cycle. Baboons are strongly attracted to members of the opposite sex independent of their immediate motivations to copulate (Collins, 1981).

Second, sex and friendship are intimately connected. During estrus, sexual relations are mediated by social bonds. Females often prefer their Friends as mates, and males are reluctant to contest access to females who share a strong bond with another male. The connection between sex and friendship extends back into the past and forward into the future: Males who show particularly intense interest in a female while she is in estrus are likely to become new Friends.

Third, the manner in which Friendships are made, the way they grow, and their persistence over time all show tremendous individual variability. There are patterns to baboon friendship, but there are few rules.

Fourth, baboons show evidence of emotions toward their Friends and mates that seem to resemble feelings that people experience in similar contexts. These emotions, rather than a compulsion to out-reproduce others, are the immediate cause of the interactions and relationships that make up the fabric of their social lives.

These results support the idea that some of the most fundamental aspects of human attachment and sexuality are shared by other animals—or, at least, by baboons. Do long-term friendships exist in other nonhuman primates as well? How does a comparative perspective on sex and friendship contribute to understanding the evolution of human behavior? These questions are considered in the final chapter.

NOTES ON STATISTICS

[1] Comparison of the number of Friend and Non-Friend dyads that consorted more often than expected. Expected values were derived

from each male's overall consort activity during the study period. The results are based on 12 estrous cycles of three females who cycled throughout most of the study (3 conception cycles and 9 nonconception cycles totalling 60 consort days). Fisher exact probability test, $N=30$, $p = .04$.

10 COMPARATIVE PERSPECTIVES

Adult female grooming adult male in three species of nonhuman primates: Japanese macaques (*top left*), EC baboons (*top right*), and chimpanzees (*bottom*). In all three species, females and males appear to develop long-term relationships based on the exchange of reciprocal benefits. (Photo of Japanese macaques by Jeffrey Kurland; photo of chimpanzees by Frans de Waal.)

INTRODUCTION

The preceding chapters have shown that most of a baboon's relaxed, friendly interactions with members of the opposite sex involve a few special friends. Friends groom often, spend time near one another, and exhibit an ease in one another's company rarely found in interactions between other males and females. Although many, perhaps most, friendships begin while the female is cycling, the bond that develops transcends this brief period of sexual activity and persists through the female's pregnancy and period of lactation. Some friendships last for years; a few may continue for the rest of the animals' lives.

In the context of these long-term bonds, females and males exchange benefits. The male defends the female and her immature offspring against danger, and he is likely to develop a long-term, protective relationship with her infant. The female, in turn, often favors the male as a mate when she resumes cycling, and she and her infant cooperate with him during tense interactions with other males. These benefits probably contribute to the participants' reproductive success, which suggests that friendship-forming tendencies have been favored by natural selection.

We have seen, however, that it is also useful to consider friendship from the point of view of the baboons themselves, treating them as sentient beings pursuing individual goals with emotional intensity and considerable intelligence. Much of what baboons do can only be understood if we grant them these capacities and assume that personal relationships, including friendships, are of central importance in their lives.

The goal of this chapter is to clarify the nature and significance of baboon friendship through comparison with male–female relationships in other primate species. Data on long-term relationships between male and female macaques are reviewed first. The next section discusses the prevalence of such long-term relationships among nonhuman primates in general. The final section considers the evolution of human male–female relationships in light of these data from nonhuman primates.

LONG-TERM, MALE–FEMALE RELATIONSHIPS IN MACAQUES

Apart from baboons, the best evidence for long-term, male–female relationships in nonhuman primates comes from Japanese macaques (*Macaca fuscata*), and rhesus macaques (*M. mulatta*). These terrestrial

Table 10.1. Body Size and Sexual Dimorphism in Olive Baboons, Japanese Macaques, and Rhesus Macaques[a]

Species	Mean adult female weight (kg)	Mean adult male weight (kg)	Female weight as a percentage of male weight
Olive baboons	12.8	24.3	53
Japanese macaques[b]	9.2	11.0	84
Rhesus macaques[b]	9.3	11.2	83

[a]Sources: Olive baboons: Eburru Cliffs troop; Japanese macaques: Sugiyama and Ohsawa (1982) and Gaulin and Sailer (1984); rhesus macaques: Rawlins *et al.* (1984).

[b]Weights are for adults 7 years old or older.

Asian monkeys are close relatives of the African baboons,[1] and their social organizations are very similar. Macaques, however, are considerably smaller than baboons and show less sexual dimorphism in body size (Table 10.1). Unlike baboons, macaques restrict mating to a few months of the year. During the breeding season, males and females form exclusive consort relationships much like those observed in baboons (Enomoto, 1974; Fedigan and Gouzoules, 1978; Takahata, 1982b; Lindburg, 1983; Chapais, 1983c).

Several recent studies have shown that among macaques, as in baboons, males and females develop long-term affiliative relationships, which I will simply refer to as friendships. Table 10.2 summarizes information on macaque friendships and compares them to friendships in savannah baboons.[2] This comparison broadens our understanding of friendship and suggests several important conclusions.

Similarities between Baboon and Macaque Friendships

The formation and maintenance of friendships and the types of interactions that occur among friends are strikingly similar in all three

[1]Macaques and baboons are grouped together in the tribe Papionini along with mandrills, gelada baboons, and mangabeys (Delson, 1980; Cronin *et al.*, 1980).

[2]It is important to note that most of the information about macaque male–female friendships comes from two studies (Takahata, 1982a,b; Chapais, 1981, 1983a,b,d). In both cases, the groups, although free-ranging, were provisioned by humans, and in the case of the Japanese macaques, provisioning resulted in an unusually large troop. It is not yet known whether the results for these groups are typical of the species under more natural circumstances, but long-term male–female associations have been reported from nonprovisioned troops of Japanese macaques (Takahata, 1982a).

species. For example, in macaques and baboons, most males and females rarely interact outside of the sexual context, and when interactions do occur, they are often uneasy and sometimes hostile. About 10–15% of the male–female dyads, however, exhibit a warm, persistent bond, characterized by grooming and frequent proximity. In all three species, a female tends to form friendships with males who are already friends with her close female relatives, and these friendships typically also include a bond between the male and the female's infant. Equally noteworthy, in both baboons and macaques, the relaxed, affectionate quality of male–female friendships resembles bonds formed among close kin (Takahata, 1982a; Chapter 9, this volume). Finally, in all three species, friends exchange benefits, as will be discussed below. These similarities suggest the possibility that long-term male–female relationships, like sexual relationships and kin relationships, will show recurrent patterns across primate species that can be explained by a few fundamental principles.

Individual Contributions to Friendship

A comparison of the three species helps to clarify the relationship between individual contributions to a friendship and the benefits received (Kummer, 1978). On a day to day basis, females in all three species seem to contribute more to friendships than do males: Females typically groom males more than males groom them, and females are primarily responsible for maintaining proximity between friends (there are a few exceptions in the rhesus macaques, discussed further below). At first glance, this pattern might suggest that females value friendships more than do males. However, a more thoughtful hypothesis would propose that what each partner puts into a relationship reflects the *particular* benefits that he or she expects to receive from that relationship. For example, if females benefit mainly from male protection against feeding interference and frequent aggression from other troop members, then it is likely that close proximity during routine foraging will be more important to females than to males. Conversely, if males benefit mainly because of increased, but periodic, sexual access to their friends (as proposed for baboons) or because of aid from females during occasional male-male fights (as proposed for rhesus macaques), then maintaining constant proximity to female friends would be less important to males.

Several testable predictions follow from this hypothesis. First, at those times when males are particularly likely to receive benefits from female friends, they should increase their responsibility for maintain-

Table 10.2. Comparison of Male–Female Friendships in Savannah Baboons, Japanese Macaques, and Rhesus Macaques[a]

Behaviors	Savannah baboons	Japanese macaques	Rhesus macaques
Descriptive aspects of friendship			
Number of adult females/adult males (sex ratio)	34/18 (1.89)[b]	84/8 (10.50)[c]	19/15 (1.27)[d]
Percentage of adult male/adult female dyads that were friends	12	13[e]	9[f]
Mean number of friendships and range	Females: 2.0 (0–5) Males: 3.4 (0–8)	Females: 1.0 (0–4) Males: 7.3 (3–10)	Females: 1.4 (1–4) Males: 1.7 (0–11)
Standard deviation in number of friendships	Females: 1.0 Males: 2.3	Females: 1.0[g] Males: 2.5	Females: 0.8[h] Males: 3.0
Male characteristics associated with greater number of friendships	Long-term residence; fully adult status	High rank; no data on residence	High rank; no data on residence
Female characteristics apparently preferred by males	Slight preference for older and higher-ranking females[i]	Preference for higher-ranking females; no data on age[j]	Strong preference for alpha female; no data on age[j]
Responsibility for maintaining proximity between friends. F = females more responsible than males; M = the reverse	F	F	In general: F; friendships with alpha female: M
Direction of grooming within friendly dyads. F = female grooms the male more often than he grooms her; M = the reverse	F	F	Friendships involving: high-ranking males: F; low-ranking males: M; alpha female: M

Relationships among females who are friends with the same male	Females often close in rank and therefore thought to be closely related	Females closely related: mothers and daughters; sisters	Females closely related: mothers and daughters; sisters
Formation of friendships	Friendships often develop out of sexual consort relationships	Same	No data
Duration of friendships	Many last 1–2 years; some for at least 6 years	Many last 1–2 years; some for at least 4–9 years	No data
Potential benefits of friendships for females			
Male proximity affords protection against disruption of activities by other troop members	Yes[k]	Yes	Yes
Male protects female against aggression from other females	Yes	Yes	Rarely
Male protection against aggression from other males:			
Male aggression toward females, including infliction of wounds, observed	Yes	Yes[l]	Yes
Defense against male aggression by friends	Yes	Yes[m]	Yes, but only by five highest-ranking males
Male–infant relations:			
Male infanticide reported in free-ranging populations	Yes	No	Yes
Males form bond with infants of female friends	Yes	Yes	Yes[n]

Table 10.2. (Continued)

Behaviors	Savannah baboons	Japanese macaques	Rhesus macaques
Proposed benefits to infant of bond with male	Grooming; occasional transport; co-feeding; protection from predation; adoption if mother dies; protection from aggression, including infanticide	Grooming; occasional transport; co-feeding; adoption if mother dies; protection from aggression by other troop members[o]	Grooming; occasional transport; adoption if mother dies; protection from aggression by other troop members[p]
Potential benefits of friendships for males			
Paternal investment: percentage of male–infant affiliative relationships in which male was a likely father based on consort records	56[q]	33[r]	Insufficient data
Male friends have increased chances of mating with female in future	Yes	No, according to Takahata, but see text for discussion	No, according to Chapais, but see text for discussion
Female friend is valuable ally increasing or stabilizing male rank	No; females do not intervene aggressively in male–male disputes	Yes; females intervene aggressively in male–male disputes[s]	Yes; females intervene aggressively in male–male disputes
Males use female friends and their infants as buffers during agonistic encounters with other males; buffers may reduce aggression toward that male	Yes	No evidence of this behavior	No evidence of this behavior

Bonds with females contribute to male integration into a new troop	Yes	Yes[t]	Yes[u]

[a]Except where indicated otherwise, sources are as follows: savannah baboons: this volume (including data from other baboon studies discussed in this volume); Japanese macaques: Takahata (1982a,b); rhesus macaques: Chapais (1981, 1983a,b,d). In contrast to savannah baboons, Japanese and rhesus macaque troops typically contain two distinct types of males: central and peripheral. Peripheral males are usually young natal males that will soon emigrate or subadult and adult males that have recently transferred into the troop. Peripheral males rarely interact with either central males or adult females, and both Chapais and Takahata excluded them from analysis. For this reason, they are also excluded from this table.

[b]Based on EC troop; the four large subadult males in EC were included.

[c]Based on the troop studied by Takahata; 20 peripheral subadult males were excluded.

[d]Based on the troop studied by Chapais; an unspecified number of peripheral males was excluded.

[e]Based on the number of pairs that were within 3 m more than 5% of the time during at least two of the five research periods (research periods were 2–3 months long and spanned 26 months).

[f]Based on the number of pairs showing frequent close proximity (<1 m), as indicated by pairs linked by heavy lines in Chapais (1983d, Fig. 10.3).

[g]Calculated from data given in Chapais (1983d, Fig. 10.3 and text); peripheral males excluded.

[h]Calculated from data given in Takahata (1982a, Table 3); peripheral males excluded.

[i]Based on characteristics of female friends of older, long-term resident males.

[j]Based on characteristics of: (1) females that have the greatest number of friends and (2) female friends of high-ranking males.

[k]Altmann (1980); Stein (1981).

[l]Tokuda (1961); Enomoto (1981).

[m]Kurland (1977).

[n]Kaufman (1967); Breuggeman (1973).

[o]Itani (1959); Alexander (1970); Kurland (1977); Hiraiwa (1981); Gouzoules (1984).

[p]Kaufmann (1965, 1967); Lindburg (1971); Breuggeman (1973); Redican (1976); Taylor *et al.* (1978); Vessey and Meikle (1984).

[q]Based on data shown in Table 8.4, this volume.

[r]Based on data shown in Gouzoules, 1984 (Table 6-3).

[s]Koyama (1970); Fedigan (1976); Gouzoules (1980).

[t]Breuggeman (1973); Kaplan (1978).

[u]Fedigan (1976).

ing proximity. This is clearly true when females are at peak estrus (Hausfater, 1975; Rasmussen, 1980; Enomoto, 1974; Chapais, 1983c). Baboon males also seek proximity to female friends when they are involved in tense encounters with rival males (Strum, 1983; Smuts, personal observation), as described in Chapter 8. Second, the degree to which females contribute to maintaining proximity should reflect their need for male protection. Data presented in Chapter 6 indicate that baboon females maintain closer proximity to male friends when they have young infants, the time when they are most vulnerable to interference from others (Altmann, 1980).[3] As a further test of this hypothesis, one could compare female roles in maintaining proximity to friends in the presence and absence of recently transferred males.

The female bias in grooming might have a similar explanation. Perhaps a female grooms a male friend in order to increase his motivation to remain close to her.[4] If so, the male should groom the female more when he is the one most likely to benefit from close proximity. This hypothesis is supported by the fact that male baboons groom females more often during consortships (Seyfarth, 1978b; Rasmussen, 1980) and during tense encounters with other males (Strum, 1983).

Another important factor that might affect each individual's contribution to a relationship is the value of a particular partner relative to other partners of the same sex. In all three species, males show much greater variability in the number of friendships per individual than do females (Table 10.2, lines 3 and 4). This is as true for the Japanese macaque troop, with more than 10 females per male, as it is for the rhesus monkey and baboon troops, with fewer than 2 females per male. These findings suggest that, in all three species, males differ greatly in their ability to obtain female friends either because of differences in

[3]We might expect macaque females, like baboon females, to spend more time with male friends after giving birth, but this hypothesis has not been evaluated quantitatively. In Takahata's troop, however, many friends showed higher proximity during the first part of the birth season than they did at other times (based on data shown in Takahata, 1982a, Table 3).

[4]Recall from Chapter 5 that when female baboons groomed with Non-Friend males (which occurs infrequently), they also groomed the male more than he groomed them. This finding is likely to require a different explanation from that of female grooming of Friends. For example, females may groom Non-Friend males in order to placate them and so reduce the chances that the unfamiliar male will threaten or attack them or their offspring.

competitive abilities and/or differences in their attractiveness to females.[5]

Both factors appear to be important. In baboons, for example, age, residency, and individual personality traits all appear to influence male attractiveness to females (see Chapters 5 and 9). In macaques, higher-ranking males have more friendships with females than do lower-ranking males, due at least in part to the effects of male competitive ability. For example, Chapais (1981) describes how a subordinate male attempting to associate with the alpha female was forced to leave her side whenever a dominant male approached. It is impossible, however, to separate the effects of attractiveness and competitive ability for two reasons. First, males can compete for friends by behaving in ways that make them more attractive (e.g., by grooming females or by associating with their infants). Second, the ability of males to provide benefits to females will sometimes vary as a function of their competitive ability against other males. For example, in rhesus macaques, only high-ranking males consistently defended females against other males (Chapais, 1983a), and we would therefore expect females, in general, to find these males most attractive.

If males usually vary more than do females in their value as friends, whether because of rank or other reasons, then males should attempt to convince females of their worth, and females should carefully discriminate among them. Qualitative evidence on the initiation of baboon friendships described in Chapter 9 suggests that this is true. In EC, during the early stages of a relationship, males often maintained proximity to females and groomed them. At this point, the male might have been attempting to demonstrate his value as a friend. In some cases, the female never reciprocated his attentions. In other cases, however, the female began to groom and follow the male. This behavior is likely to be of crucial significance because it indicates at least a tentative acceptance of the male as a friend.

Although, in general, males do seem to vary more in their value as friends than do females, Chapais' study reveals one notable exception. Female rhesus monkeys influenced male dominance relationships by directly intervening in fights between males (Chapais, 1983a,b). All females intervened, but the alpha female's interventions were partic-

[5]Differences among males in their motivation to form friendships are unlikely to account for a substantial amount of male variability in number of friends, because males without friends are often observed trying to form bonds with females (Chapais, 1981; Chapter 9, this volume).

ularly important, since other females tended to support her actions. This meant that the alpha female was considerably more valuable to males than any other females, and, as we might expect, males preferred her as a friend. These preferences were shown in three ways. First, she had more than twice as many friends as most other females. Second, her male friends bore primary responsibility for maintaining proximity, and they groomed her more than she groomed them. Third, she was the only female whom males did not attack.

At least two important factors are thus likely to influence individual contributions to a relationship. First, an individual's contribution to a relationship will reflect the particular benefits that he or she expects to receive. For example, if frequent protection by male friends is important to females, females will generally contribute more to maintaining proximity than will males. Second, what individuals put into a relationship will reflect the value of the partner relative to other members of the same sex. The more valuable an individual is compared to others, the more a potential partner will be willing to initiate and maintain his or her relationship with that individual. The alpha female rhesus monkey is so valuable as a friend that males are apparently willing to shoulder most of the responsibility for maintaining the bond.

Benefits of Friendship

Some of the benefits exchanged by macaque and baboon friends are very similar. In all three species, females are vulnerable to aggression and wounding by males, and friendship apparently affords them a measure of protection against these abuses. In baboons and Japanese macaques, females also receive protection from male friends during disputes with other females, but this was not true in the rhesus monkeys studied by Chapais. Although female macaques show aggression toward newcomer males, and female baboons normally do not (Packer and Pusey, 1979), in all three species friendships with females are crucial to male integration into a new group (Table 10.2). In other ways, however, the benefits associated with male–female relationships vary among the three species. Three differences seem particularly important:

Male–Infant Relationships. Although macaque males tend to form close bonds with the infants of their female friends, infanticide has not been reported in these species, and there is thus no evidence that male-infant bonds serve to protect infants against violent aggression from other males.[6] In addition, rhesus and Japanese macaque males generally do not carry infants during tense interactions with other males (Stein, 1981), although infant use as a "passport" to achieve proximity to other males has occasionally been observed in Japanese macaques (Itani, 1959). The fact that infanticide and infant-carrying are apparently rare in macaques provides indirect support for the hypothesis that, in baboons, infant-carrying and male protection of infants against attacks by other males are related phenomena (see Chapter 8).

Friendship and Sex. Both Takahata (1982a,b) and Chapais (1981, 1983a) claim that, unlike male baboons, male macaques do not enjoy increased opportunities to mate with their female friends. The evidence for these claims, however, is not definitive. Takahata (1982a,b) reported that friendly dyads tended to avoid mating during the breeding season: Only 8% of friends were observed mating compared to 22%

[6]It is not known why infanticide is rare or absent in macaques. In provisioned troops, many female macaques conceive regularly every breeding season whether or not their previous infants survive (e.g., Drickamer, 1974; Sugiyama and Ohsawa, 1982). For this reason, a male who kills an infant may be less likely to enjoy increased mating opportunities in macaques compared with baboons and other species that breed throughout the year . However, this does not seem to be an adequate explanation, since under more natural conditions, most female rhesus and Japanese macaques give birth only once every other year (Hiraiwa, 1981; Melnick and Pearl, 1986). Under these circumstances, infanticide could presumably hasten conception by a year. A second possible explanation is that female macaques are considerably more effective than female baboons in preventing infanticidal attacks because of reduced sexual dimorphism in body size. This hypothesis is supported by three observations. First, in contrast to baboons, in Japanese macaques female aggression can drive a male out of a troop or result in serious injury (Packer and Pusey, 1979). Second, Japanese macaque females, like baboon females, are extremely sensitive to male proximity to infants, and the faintest hint of infant distress can provoke explosive gang attacks on a nearby male even if he shows no apparent intention of harming the infant (Kurland, 1977). Third, male Japanese macaques go out of their way to avoid young infants and sometimes even show fearful responses to infant approaches (Itani, 1959; Alexander, 1970; Kurland, 1977; Hiraiwa, 1981). This circumspect treatment of infants is probably a response to the threat of female retaliation. These last two observations suggest that females fear male aggression toward infants, and it is possible that further observations of unprovisioned macaques will show that infanticide occasionally occurs. [Note added in proof: Male infanticide in wild rhesus macaques has just been reported for the first time (Ciani, 1984). The observer saw an adult male grab an infant and bite it severely on the neck and face. Six nearby females mobbed the male, and he dropped the infant and ran away. The infant died within 30 minutes.]

of nonfriend dyads, and this difference was significant (1982a, p. 14). However, data given by Takahata (1982a, p. 16, Table 10) show that in the first and second breeding seasons after a friendship was formed, 42 and 33%, respectively, of the friendly dyads were observed mating—a considerably higher percentage than the 22% of nonfriends seen mating. This suggests that for the first year or two, friendship increased a male's chances of mating with a female, but that sometime after this, a change occurred and friends began to avoid sex.[7] Takahata hints at this difference when he states, "There is little possibility that the females will bear offspring inheriting the genes of their PPR [friend] males, except for the offspring who are born within a few years after PPR-formation" (1982a, p. 20). In provisioned troops like the one Takahata studied, males often remain in the troop for many years, but in unprovisioned troops, males normally emigrate after 3–4 years of residence (Takahata, 1982a). Thus, under more natural conditions, most males will have left a troop by the time the mating inhibition between friends develops, and the absence of an association between friendship and sex may be an artifact of the unusual number of relationships persisting for more than 4 years in provisioned troops. Two other studies of Japanese macaques also reported that affiliated pairs rarely mated, but both of these studies were also conducted on provisioned troops (Enomoto, 1974; Fedigan and Gouzoules, 1978; Baxter and Fedigan, 1979).

Chapais (1981, 1983a), like Takahata, concluded that males did not enjoy increased mating opportunities with female friends. This conclusion rests on a qualitative comparison of male–female friendships (derived from grooming relationships during the birth season) and male–female consort relationships during the following breeding season. Chapais (1981) describes many cases in which males did not breed with their female friends and other cases in which males mated with females with whom they were not previously affiliated. As argued in Chapter 8, however, qualitative comparisons of this kind fail to test the hypothesis that friendship promotes mating; what is needed is a quantitative test that compares each male's mating activity with friends

[7]In some of the affiliated dyads that did not mate, Takahata observed the female avoiding male sexual initiatives (Takahata, 1982a,b); the reverse pattern was not observed. This suggests that females may be primarily responsible for low rates of mating among friendly pairs. Females are also primarily responsible for mating avoidance among close kin (Enomoto, 1974), and it is possible that after several years of intimate interaction with a male, psychological mechanisms that normally serve to promote incest avoidance are activated (Takahata, 1982a).

with that expected based on his overall mating activity. I conducted such a comparison using data presented by Chapais (1981, Table 7.15).[8] When all mating activity was considered, the five males involved in long-term, persistent grooming relationships with females all mated with a higher percentage of previous grooming partners (mean = 73%) than with females overall (mean = 42%). Unfortunately, the sample sizes were too small to attach a significance level to this finding. A second quantitative comparison was conducted using only those matings that might have resulted in fertilization. The results were equivocal: The male responsible for most (8 of 13 matings involving males who had persistent grooming relations with females) of these matings did *not* mate with his grooming partner [this was a young male who had recently risen in rank due to support from his mother, the alpha female, and his mating success seemed to be a function of his newly acquired status (Chapais, 1983a,c)]. However, of the five remaining matings, two involved grooming partners. Chapais is right to stress that prior affiliation is not a prerequisite to mating, but these results suggest that, nevertheless, such affiliations may in some cases increase male mating opportunities.

Female Intervention in Male–Male Disputes. Whether or not male macaques derive mating advantages from friendship, they do appear to receive substantial benefits from alliances with female friends. Among rhesus macaques, female intervention in male–male disputes can strongly influence male dominance relationships (Chapais, 1983d). Female interventions also appear to affect male status in Japanese macaques (Koyama, 1970), and female aggression—or lack of it—can determine the fate of a new male attempting to enter a group (Fedigan, 1976; Packer and Pusey, 1979). Although female baboons can affect the course of male disputes by cooperating with male attempts to use them as "buffers" (Strum, 1983; Chapter 8, this volume), there is no evidence that such female support has any long-term effects on male dominance relationships. As noted earlier, sexual dimorphism in body size is considerably lower in macaques than in baboons (Table 10.1), and it seems likely that this is why female aggression toward males is more common and more effective in macaques (Packer and Pusey, 1979).

[8]Persistent, long-term grooming relationships (labeled type "A" relationships by Chapais) were used as a conservative measure of male–female bonds. When measuring male mating activity, I excluded dyads that were members of the same matrilineage, since these animals rarely mate (Chapais, 1981, 1983a).

Discussion

This comparison of the benefits of friendship in macaques and baboons suggests two important conclusions. First, even among closely related species with very similar social organizations, the benefits that males and females offer one another may differ substantially due to particular characteristics of each species (e.g., degree of sexual dimorphism). In fact, it is likely that benefits will vary even within species because of demographic and ecological differences between populations. Second, despite these differences, the relationship that supports the exchange of benefits appears to be fundamentally similar in all three species. This suggests that it is the existence of reciprocity, rather than the particular benefits exchanged, that gives friendship its essential character.

MALE–FEMALE RELATIONSHIPS IN OTHER NONHUMAN PRIMATES

If reciprocal exchange of benefits indeed forms the basis for male–female friendships in primates, then such relationships should develop whenever opportunities for reciprocity exist between males and females who interact repeatedly over long periods of time. This leads to the prediction that male–female friendships will be found in many nonhuman primate species.

The evidence available to test this prediction is fragmentary. Several reports describe unusually strong bonds between particular adult males and females in various macaque species (e.g., Angst, 1975; Estrada and Sandoval, 1977), but systematic evidence for friendship exists only for rhesus and Japanese macaques. Affiliative relationships, measured by grooming, sitting close, and travelling together, apparently exist between adult male and female wild capuchin monkeys (Izawa, 1980; Robinson, 1981). Among common chimpanzees, no direct evidence indicates special relationships between unrelated adult males and females, but one study suggests that females prefer to mate with males who: (1) spend more time with estrous females, (2) groom these females more, and (3) share food with females (Tutin, 1979). If long-term, special male–female bonds do exist in common chimpanzees, they will be difficult to document because anestrous females spend most of their time alone (Wrangham and Smuts, 1980). However, among pygmy chimpanzees, where females are much more sociable, preliminary evidence indicates that particular male–female pairs tend to forage together in the same subgroup (Kano, 1982; Kitamura, 1983). Observations of woolly spider monkeys also suggest the existence of special relationships between adult males and females (K. Strier, personal communication).

Special male–female bonds also exist in species that live in one-male, multi-female groups (in the species mentioned below, the group may also contain a second male, but he usually does not breed). Harcourt (1979) found that the dominant male silverback in two groups of mountain gorillas had stronger bonds with some females than with others, and Tilford and Nadler (1978) reported the same for a group of captive lowland gorillas. Among gelada and hamadryas baboons, one female usually has a much closer relationship with the breeding male than do the other females in the unit (gelada: Dunbar, 1983; Kummer, 1975; Mori, 1979c; hamadryas: Sigg, 1980). These relationships, however, are not strictly comparable to those among savannah baboons and macaques, since the female often has only one male to interact with and since the male, in a certain sense, has a special relationship with all of the females in his unit.

Monogamous species, by definition, show long-term affiliative bonds between particular males and females, but since these bonds do not occur within the context of a larger group, they are not directly relevant to this discussion.

Discussion

The studies reviewed above suggest that special male–female relationships probably exist in a number of primate species, but firm evidence for such relationships is lacking for most nonmonogamous primates. It is not valid to assume, however, that lack of evidence for special male–female relationships indicates their absence. Baboons were studied in numerous places for over 10 years before special male–female bonds were reported (Ransom and Ransom, 1971), and several more years passed before these relationships were subjected to systematic scrutiny (Seyfarth, 1978b). Similarly, Japanese and rhesus macaques were observed intensively for 15–20 years before special male–female relationships were identified as a central feature of macaque society, although the existence of such relationships was noted earlier (Itani, 1959; Fujii, 1975; Altmann, 1962; Kaufman, 1967; Lindburg, 1971). These examples show that first-rate researchers can spend years studying a species before male–female special relationships become well documented and their importance fully recognized. The inevitable conclusion is that the frequency and significance of such relationships among primates is presently virtually unknown.

In order to rectify this gap in knowledge, it will be necessary for researchers to collect and analyze data on routine interactions between the sexes during all phases of the female reproductive cycle as a matter of course, in much the same way that they now collect and analyze data on agonistic interactions or sexual encounters. Why have primatologists not done this more often in the past? There are at least three possible reasons. First, although some aspects of social behavior can be investigated in a relatively short study, an understanding of affiliative relationships requires long-term observations of known individuals, and this has not always been possible in field studies. Second, androcentric biases have led to an overemphasis on male-male competition and an underemphasis on female choice as determinants of mate selection. Similar biases have affected many areas of primate research (Fedigan, 1982; Hrdy and Williams, 1983). A third reason is revealed by the frequency of casual references to male–female affiliations in published reports without any discussion of their meaning. This suggests that observers sometimes assume that these relationships simply reflect idiosyncratic, individual affinities that are of no general significance. This assumption reflects a subtle but pervasive neglect, until recently, of the study of long-term, individual relationships among nonhuman primates.

THE EVOLUTION OF HUMAN MALE–FEMALE RELATIONSHIPS

Introduction

There has been a great deal of speculation in recent years about the selective pressures that led to the evolution of the human pair bond and male parental care. The behavior of nonhuman primates has played a central role in many of these discussions, but, until very recently, comparative evidence that bears most directly on this issue—namely information on long-term male–female and male–infant bonds in other primates—has not been available. In its absence, investigators have developed several theoretical assumptions about the relationship between sex, parental investment, and male–female bonds in human ancestors. These assumptions, which have become widely accepted, are reconsidered below in light of new data from nonhuman primates.

Paternity Certainty and Male Parental Investment

Most recent researchers have assumed that among human ancestors male investment in infants remained minimal until males became reasonably certain that they were investing in their own offspring (e.g., Symons, 1979; Alexander and Noonan, 1979; Lovejoy, 1981; Strassman, 1981; Turke, 1984). The development of a pair-bonded mating system, in which females restricted their mating to a single male, therefore constitutes a crucial evolutionary step that allowed males, for the first time, to make a significant contribution to the welfare of mothers and infants (Symons, 1979; Alexander and Noonan, 1979; Lovejoy, 1981; Strassman, 1981; Turke, 1984).

This assumption rests on neo-Darwinian theory, which predicts that male behaviors that benefit infants at some cost to the male (male parental investment) will be directed toward a male's own infant rather than the infants of some other male (Trivers, 1972). This fundamental principle has led researchers to predict that the degree of male care of infants in nonhuman primates should reflect paternity certainty, both within groups (Hrdy, 1976; Kurland, 1977; Busse, 1984b) and across different species (Redican, 1976; Tilford and Nadler, 1978; Alexander and Noonan, 1979; Bales, 1980). This prediction is supported by the fact that the nonhuman primate species showing the greatest amounts of male parental care are generally pair-bonded (Kleiman, 1977; Whitten, 1986). Since females in these species typically mate with only one male (e.g., Leighton, 1986; Robinson *et al.*, 1986), paternity certainty is high.

Paternity certainty is also relatively high in primate species that live in multi-female groups with only one breeding male, and it has been argued that these species also should show marked male care, especially in comparison with species that live in multi-male, multi-female groups (Redican, 1976; Kleiman, 1977; Tilford and Nadler, 1978; Alexander and Noonan, 1979). The quantitative data required for a rigorous test of this hypothesis are lacking, but qualitative evidence (Table 10.3) indicates the opposite pattern: Although patterns of male care vary a great deal among multi-male species, several multi-male species show considerably more male care of infants than do any single-male species (Snowdon and Suomi, 1982; Smuts, 1982; Whitten, 1986).

One possible explanation for these results is that, in one-male groups, females mate with males other than the resident leader so often that paternity certainty is even lower than it is in multi-male species. For example, Alexander and Noonan (1979) claimed that "adulterous" matings substantially reduce paternity certainty in hamadryas baboons. The only research team to study wild hamadryas baboons, however, has never seen a female complete a copulation with a male other than "her" male during several thousand hours of observation (Sigg *et al.*, 1982). Copulations with other males are apparently more common in certain other species that live in one-male groups (e.g., hanuman langurs: Hrdy, 1977; patas monkeys: Chism *et al.*, 1982;

Table 10.3. Male–Infant Care in One-Male and Multi-Male Groups of Monkeys and Apes[a]

Species	One-male (O) or multi-male (M) groups	Male care of infants	Source
Hanuman langur[b]	O/M	–	Hrdy, 1977
Dusky leaf monkey[b]	O/M	–	Curtin, 1980
Banded leaf monkey[b]	O/M	–	Curtin, 1980
Patas monkey[c]	O	–	Hall, 1967;[d] Chism, 1978
Lowe's guenon	O	–	Bourliere *et al.*, 1970
Diana monkey	O	–	Byrne *et al.*, 1983[d]
Syke's monkey	O	–	Rowell, 1974[d]
Mountain gorilla[e]	O	+	Fossey, 1983
Gelada baboon[f]	O	Limited[g]	Mori, 1979a,b; Dunbar, 1984
Hamadryas baboon[f]	O	Limited[h]	Kummer, 1968
Blackcapped capuchin[i]	O	–	Defler, 1982
Blackcapped capuchin[i]	M	+	Izawa, 1980
White-fronted capuchin	M	+	Defler, 1982
Squirrel monkey	M	–	Vogt, 1984
Crab-eating macaque	M	–	Mitchell and Brandt, 1972[d]
Pig-tailed macaque	M	–	Mitchell and Brandt, 1972[d]
Rhesus macaque	M	Limited[j]	This volume
Japanese macaque	M	+	This volume
Barbary macaque	M	+	Taub, 1984
Stumptailed macaque	M	+	Gouzoules, 1975;[d] Estrada, 1984; Smith and Peffer-Smith, 1984[d]
Sooty mangabey	M	+	Hunter, 1978[d]
Grey-cheeked mangabey	M	+	Chalmers, 1968
Savannah baboon	M	+	This volume
Vervet monkey[k]	M	–	Struhsaker, 1967
Chimpanzee	M	–	Goodall, 1975; Nishida, 1983

[a]In most cases, only qualitative data are available on the frequency of different types of male care of infants. A minus is given when the observer reported that males rarely interacted with infants in any way or when the only frequent interactions observed were hostile ones. A plus is given when the observer reported frequent holding, nuzzling, carrying, and grooming of infants. In many of these species, observers also saw males behaving protectively toward infants in the presence of conspecifics and/or the observer. In some species, males also adopted orphaned infants (e.g., savannah baboons, rhesus and Japanese macaques, gorillas).

[b]These three Asian colobines are found in both one-male and multi-male groups. There is no evidence indicating that the amount of male care varies with the number of males in the group, but this question has not been studied in any detail.

[c]Patas monkey groups usually have only one breeding male, but on occasion, several males will temporarily enter the group and mate (Chism *et al.*, 1982). Although such multi-male influxes have not been reported for the next three species (all members of the genus *Cercopithecus*), they have been observed in the congeneric redtail monkey and blue monkey (see text).

[d]Indicates captive study.

[e]Within the same habitat, some mountain gorilla groups have only one adult male and others have more than one, including an older male and one or more younger ones. Younger males, who are probably often sons or brothers of the older male, usually do not breed (Fossey, 1983).

[f]In gelada and hamadryas baboons, the one-male units associate in larger groups, but males mate only with females from their own unit.

[g]As noted in the text, the breeding male in gelada groups rarely interacts with his infants. Most infant care is shown by young, follower males who are attempting to form bonds with females.

[h]Although hamadryas are sometimes cited as an example of a one-male species showing high amounts of paternal care (e.g., Redican, 1976; Hamilton, 1984), Kummer's (1968) description of male care in hamadryas makes it clear that carrying and grooming of infants involves mainly subadult and young adult males that have not yet bred. Extended infant care by young males is apparently related to the male's attempts to kidnap infant and juvenile females in order to form his own breeding unit. Once a male has established his own group, male carrying and grooming of infants is uncommon (Kummer, 1968, p. 65).

[i]In one habitat, blackcapped capuchins were found in one-male groups, but in another habitat, the groups tended to be multi-male.

[j]Usually, adult male rhesus monkeys rarely interact with infants (Vessey and Meikle, 1984). However, males occasionally develop strong bonds with particular infants (Breuggeman, 1973; Taylor *et al.*, 1978), and, as in Japanese macaques, males sometimes form long-term, intimate, care-taking relationships with orphans (Berman, 1982; Vessey and Meikle, 1984).

[k]Very small vervet groups sometimes have only one adult male (Struhsaker, 1967).

redtailed monkeys: Cords, 1984; blue monkeys: Tsingalia and Rowell, 1984). These copulations are generally linked to specific occurrences, such as harassment of a group by an all-male band (langurs) or invasion of a group by several solitary males (patas, blues, redtails), events that often precede replacement of the resident male by another male. In all of the species mentioned above, observers witnessed long periods of stability in between invasions or takeovers, and during these stable periods, females mated primarily with the resident male. The frequency of copulations with nonresident males and their effects on paternity certainty are not yet known, but current evidence indicates that promiscuous mating is not so common in one-male groups as it is in most multi-male groups.

If paternity certainty is the primary determinant of male investment in infants, then two other predictions follow for groups that have more than one male. First, potential fathers should show more male care in groups in which one or a few males have a relatively high probability of being true fathers and less care in groups in which several males have a roughly equal probability of fathering offspring. Data on macaques do not support this prediction. Female Barbary macaques initiate and terminate a series of brief consortships with every adult and subadult male in the group around the time of ovulation, leading to a situation in which any male could be the father, but no particular male can be considered a likely father (Taub, 1980). Yet male care of infants is considerably more elaborate in this species than in any of the other macaques (Taub, 1984), even though females in other species of macaques typically consort with only a small fraction of all potential breeders around the time of ovulation (e.g., Enomoto, 1974; Lindburg, 1983; Takahata, 1982b; Chapais, 1983c).

Second, the paternity certainty hypothesis predicts that within groups, males should care only for infants that they are likely to have fathered. In baboons and macaques, however, males are often observed caring for infants that are probably not their own (baboons: Chapter 8, this volume; Stein, 1981, 1984; macaques: Takahata, 1982a; Gouzoules, 1984). The behavior of gelada baboon males is also difficult to explain in terms of paternity certainty. The resident, breeding male rarely holds, carries, or grooms his own infants, but "follower" males—young adult males attempting to join a single-male breeding unit—engage in frequent affiliative interactions with infants they could not have fathered (Mori, 1979a,b; Dunbar and Dunbar, 1975; Dunbar, 1984). Female geladas have great influence over male group membership (Dunbar and Dunbar, 1975; Mori, 1979c), and one observer has suggested that the follower cares for an infant in order to help him to develop "a specially intimate relationship" with the mother (Mori, 1979b, p. 106).[9]

Data that allow a comparison of male care in multi-male versus one-male groups of the same species provide a further test of the paternity certainty hypothesis. Izawa (1980) reported frequent affiliative interactions between males and infants in a multi-male group of blackcapped capuchins. Defler (1982) observed two one-male groups of this species in a different habitat. In contrast to Izawa, he never saw males carry infants, and infants were not seen playing near males. However, in the white-fronted capuchin, a closely related species living in multi-male groups in the same habitat, males were extremely tolerant of infants and were seen carrying infants away from the observer (Defler, 1982).

Stein and Stacey (1981) conducted a detailed, quantitative comparison of male–infant interactions in two adjacent groups of savannah baboons. In one very small group, a single male had been the only breeding male present for many months, and he was almost certainly the father of the sole infant in the group. The other larger and more typical group contained several infants and several breeding males. Consort records were available for the conception cycles of these infants, and in most cases, the mother was observed copulating with at least two different males who were considered possible fathers of her infant. The possible fathers in the multi-male group had significantly higher scores than the certain father in the one-male group on all three measures of male care of infants: the amount of time spent in contact with infants, the frequency of male approaches to infants, and the frequency of male grooming of infants. How can these findings be explained?

Reciprocity

The data presented in this volume suggest an alternative to the paternity certainty hypothesis: Male primates care for and protect infants (and their mothers) in order to derive reciprocal benefits from females and infants. These benefits might include acceptance of an unfamiliar male into a group, female support of that male during competition with other males, use of the infant during interactions

[9]The resident gelada male shows frequent affiliative and protective behaviors toward infants only after he has been defeated by another male and is no longer able to breed (Mori, 1979b; Dunbar, 1984). One interpretation of this finding is that females will tolerate a defeated male in the group only if he continues to provide benefits to them and their infants.

with other males, or greater mating opportunities in the future. According to this "reciprocity" hypothesis, male care of infants will be most common when:

1. Infants benefit from male care. This is important because male care of infants should not influence the distribution of female benefits to males unless such care is valuable.
2. Females (and/or infants) are able to offer males substantial benefits in return for male care of infants (such as acceptance into the group, agonistic support, or mating opportunities).
3. Females have opportunities to compare the behavior of different males and then, on the basis of this comparison, distribute benefits to some males at the expense of others.

The reciprocity hypothesis is consistent with the following observations:

1. Male–infant affiliative relationships are more common in multi-male groups than in single-male groups, both within and between species (see above). This is expected for several reasons. First, in multi-male groups, females have more opportunities to compare the behavior of different males and to distribute benefits on the basis of these comparisons. Second, males might derive greater benefits from relationships with infants in multi-male groups because, unlike males in one-male groups, they can routinely use these infants to mediate interactions with other males living in the same group (Stein and Stacey, 1981). It is also possible that male bonds with particular infants contribute more to infant survival in multi-male groups than in one male groups, because in the former, infanticide can occur whenever males unlikely to be the infant's father are living in the group, whereas in one-male groups, infanticide is likely only when a strange male invades the group (Smuts, 1982).

2. In all of the studies that have considered both male–infant relationships and special relationships between males and females, males formed bonds primarily with the infants of their female friends (savannah baboons: Chapter 8, this volume; gelada baboons: Dunbar, 1984; stumptail macaques: Smith and Peffer-Smith, 1984; rhesus macaques: Kaufman, 1967; Breuggeman, 1973; Japanese macaques: Takahata, 1982a). In multi-male groups, the male's relationship with the mother is a better predictor of his relationships with infants than is paternity (Chapter 8, this volume; Berenstain *et al.*, 1981; Stein, 1981), and males often direct "paternal" care toward infants that they

are unlikely to have fathered (baboons: Chapter 8, this volume; Stein, 1981; Stein and Stacey, 1981; gelada baboons: Dunbar and Dunbar, 1975; Mori, 1979b; macaques: Gouzoules, 1984).

3. In macaques and baboons, males derive important benefits from their relationships with females and infants (baboons: Chapter 8, this volume; macaques: see above).

In summary, male care of infants is likely to arise when circumstances allow infant care to become an integral part of a mutually advantageous, reciprocal exchange of benefits between males and females. According to this hypothesis, the male does not have to be the father of an infant to benefit from its care, and male care of infants is best understood as a form of mating effort rather than as parental investment.[10] In other words, male care of infants benefits the male not because it raises the fitness of offspring already sired, but because it increases the probability that the male will father offspring in the future. Nevertheless, paternity certainty may sometimes play an important role in such a system, since, all else being equal, a male will benefit more by investing in his own infant than by investing in another male's infant. The data, however, suggest that often, all else is not equal.

This discussion calls into question the common assumption that during human evolution, male care of infants and a pair-bonded mating system were intimately linked. If male care of infants in nonhuman primates is best understood as one component of a mutually advantageous, reciprocal relationship between a male and a female, then it is likely that early hominid males made significant contributions to the welfare of mothers and infants long before the evolution of the pair bond. It is also possible that, even in modern humans, care and protection of women and children by men sometimes represents mating effort rather than parental investment.

Although this view is consistent with data from baboons and macaques, some might argue that it lacks support from studies of chimpanzees, one of our closest living relatives and a species often

[10]Reproductive effort refers to "the total resources of time and energy used by an animal in reproduction" (Krebs and Davies, 1981, p. 117). Reproductive effort can be partitioned into two parts: parental investment (care and rearing of offspring) and mating effort (acquiring mates). Since the amount of reproductive effort an animal can expend is limited, it is usually assumed that an increase in parental investment must be balanced by a decrease in mating effort and vice versa (Williams, 1966; Gadgil and Bossert, 1970; Trivers, 1972; Kurland and Gaulin, 1984).

considered to be the best model for early hominid social behavior (e.g., Goodall and Hamburg, 1975; McGrew, 1981b; Tanner, 1981). As far as is known, chimpanzees exhibit neither long-term bonds between particular males and females (other than close kin) nor frequent male care of infants. However, chimpanzees represent a rather special case, not because they are closely related to humans, but because they have a very unusual social system. The relevant results from two long-term studies (each over 20 years) in Tanzania can be summarized as follows:

1. Male chimpanzees born in the same community form a cooperative team whose joint goal is to defend an area as large as possible against incursions by males from other communities (Wrangham, 1975, 1979a, 1985; Nishida, 1979; Nishida *et al.*, 1985). Community defense can involve severe aggression between males from neighboring communities, sometimes resulting in death (Goodall *et al.*, 1979; Nishida *et al.*, 1985).

2. Most females transfer out of their natal community at adolescence into a nearby community (Pusey, 1980; Nishida, 1979; Nishida *et al.*, 1985). After transferring, females establish a core area within the larger community range (Wrangham, 1979b; Wrangham and Smuts, 1980). Females usually remain in this community for many years, often until death, and they mate primarily with the males whose range they share (Pusey, 1980; Tutin, 1980; Nishida *et al.*, 1985). Access to estrous females appears to be the main benefit that males derive from community range defense (Wrangham, 1975, 1979a; Nishida, 1979).

3. Mothers spend most of their time alone, but they occasionally associate with males in large parties gathered at abundant food sources (Wrangham and Smuts, 1980). When males and females do meet, the existence of affiliative bonds is indicated by grooming, hugging, kissing, and a wide variety of other friendly gestures (Goodall, 1975).

4. Anestrous females and infants are vulnerable to severe attacks by males from other communities (Bygott, 1972; Goodall *et al.*, 1979; Goodall, 1977, 1979). Females are particularly vulnerable to attacks (*a*) in areas near the boundary of another community's range and (*b*) when the males of their community are losing a prolonged struggle against males from an adjacent community. Successful male defense of community ranges therefore protects females from aggression by strange males.

5. Males also protect recently transferred females from aggression by resident females (Pusey, 1980). They have also been observed protecting infants against infanticide by females (Goodall, 1979).

6. Female choice of male associates (and therefore of mates) appears

to be based on the ability of a group of males to effectively defend a large community range (Nishida *et al.*, 1985). When effective defense was severely compromised as a result of a decline in male numbers, females who had associated with one community for many years deserted their males en masse and moved to the adjacent community, which had many more males and a much larger community range (Nishida *et al.*, 1985).

These facts suggest that male and female chimpanzees do establish long-term, reciprocal relationships, but in contrast to baboons and macaques, these relationships involve an association between individual females and an entire group of males. The chimpanzee data also suggest that, in the context of these long-term relationships, males make a substantial contribution to the welfare of females and infants by protecting them from strange males. Thus, although chimpanzees lack the particular type of male–female bonds and male care of infants found in baboons, both species exhibit reciprocal exchanges of benefits between adult males and females within the context of long-term, affiliative relationships. It is likely that our ancestors showed similar reciprocal arrangements, although the precise form that these arrangements took will never be known.

The Basis of Male–Female Bonds: Social or Economic Exchange?

Most authors considering the emergence of uniquely human male–female relationships place great emphasis on the division of labor and the resulting need for economic exchange between the sexes (e.g., Galdikas and Teleki, 1981; McGrew, 1981a; Tanner, 1981; Lovejoy, 1981; Lancaster and Lancaster, 1983). According to most scenarios, male and female hominids exhibited relatively undifferentiated social relationships until the division of labor and the food-sharing that it generated brought the sexes together into practical, economic arrangements that benefited both men and woman. Feminist critiques of the "man the hunter" hypothesis have rightly challenged the view that females were relegated to a dependent role in these economic exchanges (e.g., Tanner and Zihlman, 1976; Zihlman, 1978, 1981; Tanner, 1981). Few people, however, have seriously questioned the idea that changing patterns of food procurement and resulting economic exchanges were the primary reason for the emergence of highly differentiated, reciprocal bonds among ancestral males and females.

The nonhuman primate data suggest that this idea ought to be questioned. Monkeys and apes show highly differentiated, long-term relationships between males and females in the absence of any economic exchange. These relationships are based, instead, on social reciprocity (cf. Reynolds, 1981; de Waal, 1982). It seems likely that hominid males offered females important social benefits long before they were capable of big game hunting. The nonhuman primate data suggest that protection from aggression by other hominids, and, especially, by other males, were probably among the most important benefits that an affiliated male could provide. Hominid females, similarly, could have provided important benefits to males long before they were capable of carrying gathered food back to a base camp. These benefits might have included support of the male during competition with other males, and, of course, choosing that male as a mate. Among nonhuman primates, exchange of social benefits does not require a pair-bonded mating system, but it does often seem to depend on long-term, affiliative relationships between particular females and particular males (or, in the case of chimpanzees, between particular females and a small group of males). Given the nonhuman primate data, it seems likely that hominid ancestors also had such relationships.

If this hypothesis is correct, then the economic exchange engendered by a division of labor probably developed within the context of male–female relationships that had already existed for a very long time. The usual assumption is that close bonds between the sexes arose because of the need for food-sharing. It seems more likely that it was possible for male–female food-sharing to develop because females and males had already evolved the capacity for close bonds that allowed them to successfully negotiate reciprocal exchange.

These ideas are speculative, but so must be all attempts to reconstruct our past. Although the specific conclusions derived from primate research will continue to engender debate, a growing concensus among primatologists is emerging around a more general message: Our anthropoid relatives demonstrate extraordinary capacities for behavioral flexibility in the face of varying social and environmental conditions, and their social relationships approach our own in complexity and emotional valence. In accepting this message, we may lose a jealously guarded source of pride in our humanity: the conviction that only human beings can create rational societies. But this apparent loss pales in comparison to the potential gain: an opportunity to use the comparative method to formulate principles of social life that transcend the particulars of any given species or society and therefore promote a more universalistic approach to understanding our own behavior.

APPENDIXES I–XIV

Appendix I. Method of Estimating Female Ages

Adult females were assigned to one of five age classes: (1) young (approximately 6–8 years), (2) younger middle-aged (9–12 years), (3) older middle-aged (13–16 years), (4) old (17–20 years), or (5) very old (over 20) on the basis of four independent age assessments by N. Nicolson, S. Strum, Dr. Nasim Gulamhussein, a primate veterinarian from the Institute of Primate Research, Kenya, and myself. The assessments by Nicolson, Strum, and myself were based on pigmentation and texture of the perineal area, size and shape of sexual swellings, facial characteristics, and overall appearance, including body shape and condition of fur, as described by Strum and Western (1982). The assessment by the veterinarian was based on patterns of dental eruption and wear; this assessment was available for the 30 females who were trapped. For 29 of the 37 females, the age classifications of all available assessments were identical. For the remaining 9 females, available estimates never differed by more than one age class. When three of four assessments agreed, the female was placed in that class. In two cases, two estimates placed the female in one class and two in another. In these cases, the final age classification was based on the dental estimate.

Although these age estimates were somewhat subjective, they are considered sufficiently accurate to be useful for two reasons: (1) the high degree of agreement between independent age assessments; and (2) subsequent confirmation of the validity of using physical criteria to estimate female ages based on observations of females of known ages (personal observation and Strum, personal communication).

Females included as focal animal subjects (see Table 2.3) represented all age classes except very old (only one female, Athena, was judged to be very old; she was excluded from analyses because she frequently disappeared from the troop for weeks at a time). All females classified as young were also considered primiparous (i.e., they had given birth to only one offspring), based on Strum's records of female cycles and on the buttonlike appearance of their nipples prior to giving birth. The nipples of females who have nursed one or more infants tend to be much longer.

Appendix II. Adult Female Dominance Hierarchy from September 1977 through December 1978

	Loser																																							
Winner	DD	ZD	TH	PH	HH	ZI	HN	PO	AI	LE	EU	CI	AM	JO	IS	VN	ZN	LI	CC	EK	RH	CB	DL	AN	PY	MM	CG	JU	DP	IO	PS	OL	PA	ML	XA	LU	SO	AU	AT	
1 DD		19	4	15	7	12	4	7	9	7	5	1	2	10	7	3	7	6	7		5	4	4	4	16	5	5	5	10	10	4	6	3	5	2	3	4	3	10	243
2 ZD			8	31	7	7	6	14	9	14	10	6	3	7	2	2	12	17	4		6	14	10	7	15	11	15	11	13	9	6	13	1	6	7	10	5	4	11	340
3 TH[a]				9	2			3	2	4	7	2	2	5	2	1	2	3	1		4	2		3	1	10	3	3	3	8	2	7	2	1	3	2	1	2	3	106
4 PH[a]					6	8	6	8	19	14	7	2	8	13	2	1	7	13	7		16	12	4	6	9	8	5	6	3	13	8	2	2	3	5	9	9	4	3	250
5 HH						8	2	8	6	8	20	4	1	4	3	2	13	8	2	1	3	7	8		3	10	11	4	1	6	7	7	5	3	3	5	4		8	187
6 ZI							1	10	8	8	12	8	2	6	5		3	6	9	1	3	8	1	5	8	7	4	5	3	3	2	3	3	7	9	5	9	2	3	172
7 HN[a]				1				12	6	16	11	3	5	5	2		2	8	7		2	2	4	1		2	10	1	2	6	4	4		5	9	7	7		4	151
8 PO				1					4	10	15	5	1	10	8	2	11	8	9		11	4	9	5	10	8	9	6	5	14	7	5	1	1	8	3	4	3	1	202
9 AI										9	3	4	2	1	2	2	3	4	3		2	2	2	3	2	5	7	5	1	6	2	1	1		3	4	2	2	6	92
10 LE	1										15	9	4	9	3	3	8	17	4	2	7	12	9	5	10	13	16	12	8	11	7	16	3	11	2	5	3	15	11	256
11 EU												11	1	13	12	5	8	8	7		11	14	1	2	7	11	7	7	4	8	2	3	1	5	5	16	9	1	6	194
12 CI														19	6		5	5	8		4	6	2	3	11	3	5	6	3	4	1	1	1	3	5	5	9	2	9	128
13 AM[a]											1			1	3		6	6	1		11	4	9	2	8	6	3	7	6	10	2	13	3	6	3	4	4	5	10	141
14 JO								1							5	1	8	4	7		2	12	6	2	6	3	5	4	4	7	4	4	3	2	3	4	1	1	5	105
15 IS																	2	1	1		3	1	3	3		2	1	3	1	1	1	1	3	1	4	4	1		2	43
16 VN[a]																	3	3	2			2	1	2	4		3	5	1	3	1		1	2		1		1	1	37
17 ZN											1							1	4	1	1	5	4	1	2	3	3	4	6	1		9		3	2	3	4	1	2	64
18 LI																			13	1	10	15	6	4	12	7	8	7	9	10	5	11	16	9	10	12	11	2	1	183
19 CC																				2	2	2	1	3	7	7	6	4	2	8	4	4	2	7	3	7	2		5	78
20 EK[a]																					3	1	2		2		5	5	2	1		4	2	2	2				1	32
21 RH																						7	4		8		8	6	3	9	1	3	1	3	2	6	3	1	1	68
22 CB													1				1				1		4	3	10	5	6	8	13	6	5	11	2	6	2	11	9	3	8	120
23 DL[a]				1				1													1	1		6	19	12	7	5	5	11			5	17	4	4	15	4	4	122
24 AN																									1		1	5		2	1	1	1	1		1	1		1	17
25 PY								1													1		3			4	6	6	6	4	3	2	1	3	3	3	1	1		48
26 MM																							1				5	3	5	5	2	9	1	4	6	2		3	2	51
27 CG																												17	6	8	3	11	2	3	8	5	3	4	4	76
28 JU																		1												6	6	3	5	5	8	10	7	3	3	63
29 DP																		2									1			3	5		2	4	2	6	4	1	4	34
30 IO																		1				1									2	4	2	7	6	4	5	3	3	38
31 PS																					1											7	1	3	3	3	1	10	2	32
32 OL																									1								1	10	6	2	3	9	2	34
33 PA																																		4	6	4	4	1	4	23
34 ML																									1										1	6	10	3	1	22
35 XA																									1		1					1				7	1	4	1	17
36 LU																																					1		5	6
37 SO																																				1			3	4
38 AU[a]																																								0
39 AT																																								0
Total																																								3779

[a]Adolescent females.

Appendix III. Behaviors Recorded Continuously During Focal Samples of Adult Females

The methods used to record social interactions continuously throughout focal samples are described below. Only those methods pertinent to interactions between males and females are described here. Additional behavior categories were used to record social interactions between females (Smuts, in preparation).

"Adults" below includes all adult and adolescent females and all adult and subadult males (see Chapter 2 for definitions).

I. *Approaches and leaves over distances of 1 m*

Whenever an adult moved to within 1 m of the subject, or the subject moved to within 1 m of another adult, and the movement was followed by either (*a*) a pause by the approacher or (*b*) an avoid by the approachee, the movement was scored as an approach. The activity of the approached animal at the time the approach was initiated was scored as 1 of the 13 activity states listed in Appendix V. Instances where one animal moved past another within 1 m but did not stop moving and where the other animal did not move away were scored as "passes by" and tallied separately from approaches.

Whenever an adult moved more than 1 m away from the subject, or the subject moved more than 1 m away from another adult, a leave was scored. Occasionally, two individuals approached or left each other simultaneously, in which case a mutual approach or mutual leave was scored.

II. *Approaches and leaves over distances greater than 1 m*

Whenever proximity between the subject and an adult male changed from one of the following three distance categories to another, I noted the individual responsible for the change: distance category 2: 1–2 m; distance category 3: 2–5 m; distance category 4: >5 m. Occasionally, proximity changed because both the male and the female moved at the same time, in which case the change was scored as mutual.

III. *Interactions during or following an approach to within 1 m*

A. *Avoid.* The approached animal moved away in response to the other's approach. Avoids over distances of more than 2 m were scored only if the avoiding animal had glanced at the approacher. Avoids over distances greater than 15 m could not be reliably recorded but were noted anyway; they were tallied separately from other avoids.

B. *Supplant.* Same as above except the approaching animal took over the resting/feeding site of the animal who moved away. Avoids and supplants were rarely preceded by threats or chases by the approacher, and, when they were, the approacher was always a male. Such interactions were scored as aggressive encounters and were tallied separately from other approaches and leaves.

C. *Visual responses*

1. Look at. Prolonged gaze (at least 2 sec) in nonagonistic context. Often the approached animal watched the other approach.
2. Glance at. A quick glance at and then away from another.
3. Nervous glance. Repeated quick glances at another.

D. *Vocal responses and facial expressions.* Males and females often grunted, or made the "come-hither" face while approaching or being approached. In the "come-hither" face, the ears are flattened back against the skull and the skin around the eyes is drawn back and held taut (see Figure 1.1). Usually this expression is accompanied by lip-smacking and/or grunting. In an exaggerated form of the "come-hither" face (usually given by males), the chin is pulled in and the shoulders are hunched down and forward. Strum named this expression "come-hither" because it often seems to function as an invitation to increase proximity. Ransom (1981) calls the same expression "ears back, eyes narrowed," and Packer (1979a) refers to it as "narrowed eyes, ears flattened."

E. *Presenting and responses to presenting*

1. Present. The female presents her hindquarters to the male. Usually females placed their perineums close to the male's face, but when nervous they might present at an angle or back up toward the male, sometimes stopping up to a few meters away.

2. Responses to a sexual present.

a. Ignore. Males often ignored the female's present. They sometimes actively ignored her by gazing elsewhere or grooming themselves. At other times, they looked at her but did nothing. Both responses were scored as an ignore.

b. Vocalize. The male lip-smacks and/or grunts in response.

c. Grasp haunches. The male places one or both hands on the female's hips and holds her for a few seconds.

d. Inspect, tactile. The male touches the female's perineum. Sometimes a male placed a finger in the vagina and then sniffed his hand.

e. Inspect, olfactory. The male sniffs the female's perineum.

f. Groom. The male grooms the female's perineal area. Usually, grooming was perfunctory.

g. Sexual response. The male mounts the female. Sometimes the mount included thrusting, with or without intromission.

3. Groom present. The male stands immediately next to the female, usually laterally, or the male sits or lies near the female and presents a part of his body for grooming. These postures were usually held for several seconds and could be distinguished from standard resting postures by the closeness of the male and the artificial way he held his body. Females also presented for grooming occasionally.

a. Groom the animal who is presenting.

b. Fail to groom the presenting animal. Failure to groom was often followed by the presenting animal grooming the other.

Grooming could also follow an approach without an intervening groom present.

F. *Remaining in proximity*

1. Rest near. Resting within 1 m of the other.
2. Feed near. Feeding within 1 m.
3. Drink near. Drinking within 1 m.
4. Travel near. Travel within 1 m of the other.

G. *Submission.* Females sometimes responded to the approach of a male, or when approaching a male themselves, with submissive gestures in the absence of any aggression by the male. The following submissive gestures were recorded.

1. Fear grin
2. Fear geck

3. Scream
4. Tail-up
5. Lean away
6. Crouch
7. Startle away. The female jumps away from the male but remains within 1 m of him.
8. Nervous glance (as defined above).

H. *Greetings*

1. Greet. One animal touches the other, usually on the torso. This category excludes touching following a sexual present and the touching involved in grooming. Tactile greetings between males and females were extremely rare (they are common between females), and when they did occur they usually involved the male touching the female at the same time that he touched an infant clinging to her body.
2. Sniff mouth. The approached animal or the approacher sometimes sniffed the mouth of the other (a common interaction within and between all age/sex classes except between adult males). The other either ignored the sniff or averted his/her face.

I. *Interactions involving the female's infant*

1. Peer at infant and grunt and/or lip-smack at it.
2. Same as above, except also gently touch the infant.
3. Pull on the infant.
4. Pull infant off the mother, or pull the infant from nearby and hold it against the body.
5. Pull the infant off the mother or from nearby and carry it away. This category almost always involved an interaction with another male and was scored as a "male–infant carrying" interaction (see Chapter 8). Whenever such an interaction occurred, I recorded the sequence of events in as much detail as possible.
6. Infant approaches male.
7. Infant makes contact with male.
8. Infant climbs on male.

J. *Other*

1. Sometimes the approach of a male to a female, or vice-versa, involved no interactions between them, but the approacher interacted with a third party nearby. These were scored as "non-interactive, third party" approaches.
2. Occasionally the approacher would pause briefly after approaching, look around, and then leave without interacting with anyone. These were scored as "approach–nothing happens."

IV. *Aggression and Submission*

A. *Aggressive behaviors*

1. Stare
2. Bite
3. Chase
4. Threaten with raised eyelids
5. Threaten by slapping the ground
6. Threaten by bobbing the head
7. Attack
8. Pin to the ground while attacking

9. Hit
10. Lunge
11. Pant-grunt (aggressive vocalization). Whenever a male exhibited aggression toward a female, I recorded the context of the interaction, including the presumed cause if it could be determined.

B. *Submissive behaviors*. The female's response to aggression by a male was recorded. Categories of response included all submissive gestures listed above (III, G), plus the following:
1. Run away.
2. Solicit aid from another. The soliciting animal stood with tail raised, screaming and glancing rapidly back and forth between a potential ally and the aggressor.
3. Counter chase. A female chased a male, usually screaming and with her tail raised.
4. Mob. The female and others joined to chase the male, uttering both screams and pant-grunts (aggressive vocalizations).
5. Run to another male for protection. In these instances, this male's reaction was recorded.
6. Fight back. This category was extremely rare, but a few times I saw a female face her male aggressor, fencing with her hands while pant-grunting with open mouth.

C. *Submission over long distances*. Females sometimes showed submissive responses to males who were not approaching them and whom they were not approaching. The following categories were scored.
1. Long distance avoid. The female sees a male moving in her direction when he is over 5 m away and, after glancing at him (usually nervous glancing), she moves rapidly away. If I was not sure that the female was moving in response to a particular male, the behavior was not scored, and it should be emphasized that some long-distance avoids were missed.
2. Tail up. The female looks at a male who is at least 5 m away and raises her tail.
3. Hurry past. The female travels past a male (anywhere from 1 to 15 m away) and as she passes, she glances nervously at him and speeds up her pace. Often accompanied by tail up, fear grin, or fear geck.

V. *Context*

Whenever possible, I recorded the context of any male–female interaction. Context included details about the pair's current or prior activities or interactions that seemed relevant (e.g., interactions with others, whether the male was in consort or following a consort pair, the existence of hunting or meat-eating in the vicinity).

Appendix IV. Number of Focal Samples (FS) and Amount of Focal Sampling Time in Minutes (M) on Females at Different Reproductive Stages

Female	Total		Cycling		Pregnant		Lactating		Pregnant and lactating	
	FS	M	FS	M	FS	M	FS	M	FS	M
AI	48	1424	0	0	0	0	48	1424	48	1424
AT	48	1167	3	45	0	0	45	1122	45	1122
AU	41	990	0	0	0	0	41	990	41	990
CB	94	2194	16	480	23	374	55	1340	78	1714
CC	49	1371	16	385	0	0	33	986	33	986
CG	89	2178	11	330	23	458	55	1390	78	1848
CI	48	1413	0	0	7	210	41	1203	48	1413
DD	89	2017	16	474	24	359	49	1184	73	1543
DL	64	1689	0	0	0	0	64	1689	64	1689
DP	89	2340	7	210	21	510	61	1620	82	2130
EU	55	1590	0	0	20	600	35	990	55	1590
HH	47	1050	30	640	13	320	4	90	13	320
IO	48	1401	0	0	0	0	48	1401	48	1401
IS	48	1403	0	0	1	30	47	1373	48	1403
JO	50	1483	0	0	0	0	50	1483	50	1483
JU	49	1417	0	0	7	210	42	1207	49	1417
LE	90	2359	9	266	20	494	61	1599	81	2093
LI	48	1404	35	1050	2	30	11	324	13	354
LU	49	1435	0	0	0	0	49	1435	49	1435
ML	74	1668	16	476	24	370	34	822	58	1192
MM	66	1839	1	30	22	586	43	1223	65	1809
OL	90	2060	24	719	25	374	41	967	66	1341
PA	58	1604	0	0	24	668	34	936	58	1604
PH	67	1769	0	0	0	0	67	1769	67	1769
PO	87	2283	8	234	19	465	60	1584	79	2049
PS	49	1150	49	1150	0	0	0	0	0	0
PY	86	2277	2	60	19	480	65	1737	80	2098
RH	67	1873	0	0	23	640	44	1233	67	1873
SO	86	2249	9	260	18	449	59	1540	77	1989
XA	62	1725	0	0	25	685	37	1040	62	1725
ZD	101	2429	10	293	26	465	65	1671	91	2136
ZI	48	1371	0	0	7	201	41	1170	48	1371
ZN	43	956	38	880	5	76	0	0	5	76
Total	2127	55,578	300	7982	398	9054	1429	38,542	1819	47,387
Hours		926.3		133.0		150.9		642.4		789.8

Appendix V. Activity Categories Used in Female Focal Samples

1. *Feed standing.* Included normal standing posture or bipedal stance. Feeding was defined as ingesting food or being in the act of picking/digging food. Searching for food, climbing a tree to reach food, and chewing while travelling were not included.

2. *Feed sitting.* Normal sitting posture.

3. *Feed in a tree.* Same definition of feeding as above, but subject was off the ground in a tree or bush.

4. *Travel.* Any locomotion other than that involved during the act of feeding as defined above; subject has not fed at any time in the preceding minute.

5. *Travel/feed.* As above, except subject has fed within the previous minute.

6. *Rest standing.* Subject is standing and is not feeding or engaged in a social interaction.

7. *Rest sitting.* As above but subject is sitting.

8. *Rest lying down.* As above but subject is lying down.

9. *Groom.* Subject is grooming another animal.

10. *Being groomed.* Subject is being groomed by another animal.

11. *Self-groom.* Subject is grooming herself.

12. *Social interaction.* Subject is engaged in a social interaction other than grooming.

13. *Bad observation.* Unable to determine subject's activity.

Appendix VI. Determination of Proximity Score Weighting Factors[a]

Distance category	Distance category limits (m)	Midpoint of inner and outer limits	Reciprocal of midpoint
1	0–1	0.5	2.0
2	1–2	1.5	0.6667
3	2–5	3.5	0.2857
4	5–15	10.0	0.1

[a]Several sets of values were tried as weighting factors for a subset of six females. All produced similar results: For all six females, most males had similar, low composite proximity scores and one, two, or three had much higher scores. The values used to weight proximity scores in the final analysis were arbitrary but logical: The reciprocals of the midpoint between the inner and outer limits of the distance category were used. These values are shown in the last column of the table.

Appendix VII. Method of Calculating Composite Proximity Scores (*C* Score)[a]

		Male: SK		Male: HM	
Distance category	Weighting factor	Score	Score times weighting factor	Score	Score times weighting factor
1	2.0	1.52	3.04	0.30	0.60
2	0.6667	1.83	1.22	0.30	0.20
3	0.2857	3.36	0.96	1.22	0.35
4	0.1	8.33	0.83	1.28	0.13
C score			6.05		1.28

[a]The method of calculating the composite proximity score is illustrated using scores for the female AI with two males, SK and HM. The score is the percentage of point samples in which the male was present at a given distance. Each score is multiplied by the weighting factor for that distance (see Appendix VI). Then all four weighted scores are added together to produce the male's *C* score.

Appendix VIII. Identification of Putative Juvenile Offspring

With the exception of older infants still occasionally seen nursing in the first few weeks of the study period, maternity was not known for any EC juveniles. Based on proximity, frequency of grooming, choice of sleeping partners, patterns of defense, and relationships between juveniles themselves, all but seven of the immature animals were assigned putative mothers (these seven had no strong bonds with any adult female old enough to be a mother, and their mothers were presumed dead). With the exception of these seven, there was not a single instance where the above measures suggested more than one candidate for motherhood, and Nicolson and I agreed on all assignments. The validity of this method of identifying mothers was confirmed by a quantitative analysis of relationships between EC juveniles and *known* mothers conducted in 1981 (Wilson, 1982). It was also supported by my own observations of relationships between EC juveniles and known mothers conducted in 1983.

Appendix IX. Characteristics of Friends of Males of Different Age/Residence Categories[a]

Male age/ residence category	Male	Number of Friends	Names of Friends	Age class of Friend	Dominance rank of Friend
Long-term resident	AG	4	AI	2	9
			LE	2	10
			EU	3	11
			AT	3	37
	BZ	6	PH	1	4
			HH	2	5
			LE	2	10
			RH	2	19
			IO	1	28
			ML	2	32
	CY	4	DD	3	1
			ZI	3	6
			PO	2	8
			CC	3	18
	HC	6	ZD	3	2
			EU	3	11
			CI	3	12
			JO	3	14
			IS	4	15
			AT	3	37
	VR	6	EU	3	11
			CI	3	12
			IS	4	15
			LI	4	17
			PA	4	31
			LU	3	34
Short-term resident	AC	5	LE	2	10
			MM	3	24
			CG	2	25
			IO	1	28
			OL	2	30
	HD	5	DD	3	1
			LI	4	17
			CB	3	20
			PY	3	23
			DP	3	27

Appendix IX (continued)

Male age/ residence category	Male	Number of Friends	Names of Friends	Age class of Friend	Dominance rank of Friend
	IA	1	CG	2	25
	SK	8	HH	2	5
			ZI	3	6
			PO	2	8
			AI	2	9
			CB	3	20
			CG	2	25
			JU	3	26
			XA	2	33
Newcomer	AA	0			
	JS	0			
	TN	0			
Young adult natal male	AO	3	ZD	3	2
			PH	1	4
			MM[b]	3	24
	HM	2	DL	1	17
			LI[b]	4	21
Subadult natal male	AS	6	LE	2	10
			CG	2	25
			IO	1	28
			OL	2	30
			ML	2	32
			AU	1	36
	HS	1	AT[b]	3	37
	PL	2	LE	2	10
			IO	1	28
	PX	3	DL	1	21
			DP[b]	3	27
			ML	2	32

[a]See Chapter 1 for definitions of male age/residence classes. Only adult female Friends are listed. Based on grooming records, the following pairs of friendships involving adolescent females were identified: TH–AA, AM–AO, AM–BZ.

[b]Female was the putative mother of her male Friend.

Appendix X. Restrictions Applied to Scoring of Approaches and Leaves

I. *Approaches and leaves over distances of 1–5 m*

In general, each individual received a score each time she or he made an approach or leave, as defined in the text. There were three exceptions to this rule. First, when one individual moved into and then out of one or more of the four distance categories without pausing, an approach/leave was not scored. In other words, individuals who were "just passing by" were not counted. Second, if the same individual made two movements in the same direction in sequence without an intervening movement by either individual in the opposite direction, the two movements made by the same individual were scored as one movement only. The purpose of this restriction was to avoid the inflation of sample sizes that would result from treating one long "movement," interrupted by brief pauses to feed or interact with a third party, as multiple movements. For this reason, the number of approaches and leaves between a given pair was not always identical, but it was always similar. Third, movements that were mutual (both partners moved at the same time) were excluded from analysis; for all females such movements contributed less than 10% of the total.

II. *Approaches and leaves over distances of 0–1 m ("close" approaches and leaves)*

Due to the relatively small sample sizes for movements into and out of 1 m's proximity, biases might result from inclusion of approaches that were not paired with a leave or leaves not paired with an approach. Therefore, unpaired movements (i.e., movements occurring at the start or end of a sample or movements missed due to gaps in observation during a sample) were eliminated from analysis; the number of approaches was therefore always equal to the number of leaves within each dyad.

Appendix XI. Percentage of Time Anestrous Females Spent in Close Proximity (within 1 m) to Friends and Non-Friends[a]

Female	Friends	Non-Friend males
AI	1.5	0
AT	1.3	0.1
AU	4.2	0.1
CB	0.9	0
CC	0.9	0.1
CG(1)[b]	3.3	0.2
CG(2)[b]	1.7	0.1
CI	0.5	0
DD	1.5	0
DL	1.9	0.1
DP	1.2	0.1
EU	1.4	0.1
HH	0.5	0.3
IO	1.5	0
IS	1.1	0.1
JO	2.9	0.1
JU	2.4	0.1
LE(1)[b]	0.7	0.1
LE(2)[b]	0.8	0.1
LI	2.0	0
LU	2.7	0.1
ML	1.4	0
MM	1.0	0
OL	1.5	0.1
PA	1.6	0.1
PH	1.4	0.1
PO	4.1	0.1
PY	3.3	0.1
RH	5.8	0.1
XA	1.5	0.1
ZD	2.2	0.1
ZI	5.0	0.1

[a]Values for Friends represent the mean value for all of a female's Friends, and values for Non-Friend males represent the mean value for all Non-Friend males.

[b]Values are given for periods one and two (before and after Friends IA and AG left the troop.)

Appendix XII. Frequency with Which a Close Approach by the Male was Followed within 5 Seconds by a Leave by the Male ("Immediate Leave") for Friends and Non-Friends[a]

	Male leaves within 5 seconds	Male remains for more than 5 seconds
Friends	18	131
Non-Friends	51	25

[a]Based on focal samples of 29 anestrous females. $\chi^2 = 77.7$ (see Chapter 5, Statistics [17]).

Appendix XIII. Frequency with Which a Close Approach by the Female was Followed within 5 Seconds by a Leave by the Female ("Immediate Leave") for Friends and Non-Friends[a]

	Female leaves within 5 seconds	Female remains for more than 5 seconds
Friends	12	121
Non-Friends	33	6

[a]Based on focal samples of 29 anestrous females. $\chi^2 = 85.33$ (see Chapter 5, Statistics [18]).

Appendix XIV. Males Frequently Observed Near Infants (within 5 m)[a]

Mother	Males found near infant	Percentage of point samples in which male was near infant out of all samples in which any male was near infant: Mother absent	Mother present	Number of point samples in which any male was near infant: Mother absent	Mother present
AI	SK[b]	80	25	5	12
CC	*CY*[b]	82	50	22	4
CI	HC[b]	80	50	5	4
DL	SC[c]	60	0	5	8
DP	PX	25	17	8	12
EU	HC[b]	67	0	6	10
IO	BZ[b]	46	—	13	1
	AS[b]	23	—	13	1
IS	VR	40	50	5	4
	HC	40	25	5	4
JO	HC[b]	57	67	3	7
JU	AC[d]	43	60	7	5
LE	AC	100	30	2	10
LI	HM[b]	75	8	4	12
	HD	25	33	4	12
LU	VR[b]	75	13	4	8
MM	AC	50	27	2	11
	BZ[d]	50	18	2	11
PA	VR	50	57	2	7
PH	AO	50	0	6	2
PY	HD[b]	78	67	3	9
RH	BZ	50	68	2	19
XA	SK[b]	78	33	9	9
ZI	CY[b]	38	40	13	10
	SK[b]	31	30	13	10

[a]Data were derived from focal infant samples conducted by N. Nicolson. Point samples were taken once, at the start of each focal sample. Mothers were "absent" when they were >5 m from their infants, and "present" when they were <5 m from their infants.

[b]These males were also found within 2 m of the infant on at least 20% of all intervals in which any male was within 2 m of the infant when the mother was absent. Note that only Friends of the mother were seen near infants under these more stringent conditions.

[c]SC was a large juvenile male thought to be the brother of DL. He was the only male, other than small juvenile males, observed near DL's infant when DL was absent.

[d]Indicates that the male was not a Friend of the mother.

BIBLIOGRAPHY

Abegglen, J.J. (1984). "On Socialization in Hamadryas Baboons. A Field Study." Associated University Presses, Cranbury, N.J.
Alexander, B.K. (1970). Parental behavior of adult male Japanese monkeys. *Behaviour* **36**, 270–285.
Alexander, R.A. and Noonan, K.M. (1979). Concealment of ovulation, parental care, and human social evolution. *In* "Evolutionary Biology and Human Social Behavior: An Anthropological Perspective" (N.A. Chagnon and W. Irons, eds.), pp. 430–461. Duxbury Press, North Scituate, Mass.
Altmann, J. (1974). Observational study of behaviour: Sampling methods. *Behaviour* **48**, 1–41.
Altmann, J. (1980). "Baboon Mothers and Infants." Harvard University Press, Cambridge.
Altmann, J., Altmann, S.A., Hausfater, G., and McCuskey, S.A. (1977). Life history of yellow baboons: Physical development, reproductive parameters and infant mortality. *Primates* **18**, 315–330.
Altmann, S.A. (1962). A field study of the sociobiology of rhesus monkeys. *Ann. N.Y. Acad. Sci.* **102**, 338–435.
Altmann, S.A. (1970). The pregnancy sign in savannah baboons. *Lab. Animal Dig.* **6**, 6–10.
Altmann, S.A. and Altmann, J. (1970). "Baboon Ecology: African Field Research." University of Chicago Press, Chicago.
Andelman, S.J. (1985). Ecology and reproductive strategies of vervet monkeys (*Cercopithecus aethiops*) in Amboseli National Park, Kenya. Ph.D. Thesis, University of Washington.
Anderson, C.M. (1983). Levels of social organization and male-female bonding in the genus *Papio. Amer. J. Phys. Anthropol.* **60**, 15–22.
Angst, W. (1975). Basic data and concepts on the social organization of *Macaca fascicularis. In* "Primate Behavior: Developments in Field and Laboratory Research" (L.A. Rosenblum, ed.), pp. 325–388. Academic Press, New York.
Appleby, M.C. (1983). The probability of linearity in hierarchies. *Animal Behav.* **31**, 600–608.
Axelrod, R. and Hamilton, W.D. (1981). The evolution of cooperation. *Science* **211**, 1390–1396.
Bachmann, C. and Kummer, H. (1980). Male assessment of female choice in hamadryas baboons. *Behav. Ecol. Sociobiol.* **6**, 315–321.
Bales, K.B. (1980). Cumulative scaling of paternalistic behavior in primates. *Amer. Natur.* **116**, 454–461.
Bateman, A.J. (1948). Intra-sexual selection in *Drosophila. Heredity* **2**, 349–368.
Baxter, J. and Fedigan, L.M. (1979). Grooming and consort partner selection in a troop of Japanese monkeys. *Arch. Sexual Behav.* **8**, 445–458.
Berenstain, L., Rodman, P.S., and Smith, D.G. (1981). Social relations between fathers and offspring in a captive group of rhesus monkeys (*Macaca mulatta*). *Animal Behav.* **29**, 1057–1063.
Berenstain, L. and Wade, T.D. (1983). Intrasexual selection and male mating strategies in baboons and macaques. *Intern. J. Primatol.* **4**, 201–235.
Berger, J. (1983). Induced abortion and social factors in wild horses. *Nature (London)* **303**, 59–61.
Berman, C. (1980). Early agonistic experience and rank acquisition among free-ranging infant rhesus monkeys. *Inter. J. Primatol.* **1**, 153–170.

Berman, C. (1982). The social development of an orphaned rhesus infant on Cayo Santiago: Male care, foster mother-orphan interaction and peer interaction. *Amer. J. Primatol.* **3**, 131–141.

Bourliere, F., Hunkeler, C., and Bertrand, M. (1970). Ecology and behavior of Lowe's guenon (*Cercopithecus campbelli Lowei*) in the Ivory Coast. *In* "The Old World Monkeys," (J. Napier and P. Napier, eds.), pp. 297–333. Academic Press, London.

Bramblett, C.A., Bramblett, S.S., Bishop, D.A., and Coehlo, A.M. (1982). Longitudinal stability in adult status hierarchies among vervet monkeys (*Cercopithecus aethiops*). *Amer. J. Primatol.* **2**, 43–52.

Breuggeman, J.A. (1973). Parental care in a group of free-ranging rhesus monkeys (*Macaca mulatta*). *Folia Primatol.* **20**, 178–210.

Busse, C.D. (1981). Infanticide and parental care by male chacma baboons, *Papio ursinus*. Ph.D. Thesis, University of California, Davis.

Busse, C.D. (1984a). Tail raising by baboon mothers toward immigrant males. *Amer. J. Physical Anthropol.* **64**, 255–262.

Busse, C.D. (1984b). Triadic interactions among male and infant chacma baboons. *In* "Primate Paternalism" (D.M. Taub, ed.), pp. 186–212. Van Nostrand Reinhold, New York.

Busse, C.D. and Hamilton, W.J., III. (1981). Infant carrying by male chacma baboons. *Science* **212**, 1281–1283.

Bygott, D. (1972). Cannibalism among wild chimpanzees. *Nature (London)* **238**, 410–411.

Byrne, R.W., Conning, A.M., and Young, J. (1983). Social relationships in a captive group of Diana monkeys (*Cercopithecus diana*). *Primates* **24**, 360–370.

Carpenter, C.R. (1963). Societies of monkeys and apes. *In* "Primate Social Behavior" (C.H. Southwick, ed.), pp. 24–51. Van Nostrand Reinhold, New York.

Caryl, P. (1979). Communication by agonistic displays: What can games theory contribute to ethology? *Behaviour* **68**, 136–169.

Chalmers, N.R. (1968). The social behavior of free-living mangabeys in Uganda. *Folia Primatol.* **8**, 263–281.

Chance, M.R.A., Emory, G.R., and Payne, R.G. (1977). Status referents in long-tailed macaques (*Macaca fascicularis*): Precursors and effects of a female rebellion. *Primates* **18**, 611–632.

Chapais, B. (1981). The adaptiveness of social relationships among adult rhesus monkeys. Ph.D. Thesis, University of Cambridge.

Chapais, B. (1983a) Adaptive aspects of social relationships among adult rhesus monkeys. *In* "Primate Social Relationships: An Integrated Approach" (R.A. Hinde, ed.), pp. 286–289. Blackwell, Oxford.

Chapais, B. (1983b) Matriline membership and male rhesus reaching high ranks in their natal troops. *In* "Primate Social Relationships: An Integrated Approach" (R.A. Hinde, ed.), pp. 171–175. Blackwell, Oxford.

Chapais, B. (1983c). Reproductive activity in relation to male dominance and the likelihood of ovulation in rhesus monkeys. *Behav. Ecol. Sociobiol.* **12**, 215–228.

Chapais, B. (1983d). Structure of the birth season relationship among adult male and female rhesus monkeys. *In* "Primate Social Relationships: An Integrated Approach" (R.A. Hinde, ed.), pp. 200–208. Blackwell, Oxford.

Cheney, D.L. (1977a). The acquisition of rank and the development of reciprocal alliances among free-ranging immature baboons. *Behav. Ecol. Sociobiol.* **2**, 303–318.

Cheney, D.L. (1977b). Social development of immature male and female baboons. Ph.D. Thesis, University of Cambridge.

Cheney, D.L. (1978). Interactions of immature male and female baboons with adult females. *Animal Behav.* **26**, 389–408.

Cheney, D.L., Lee, P.C., and Seyfarth, R.M. (1981). Behavioral correlates of non-random mortality among free-ranging female vervet monkeys. *Behav. Ecol. Sociobiol.* **9**, 153–161.

Cheney, D.L., and Seyfarth, R.M. (1977). Behavior of adult and immature male baboons during inter-group encounters. *Nature (London)* **269**, 404–406.

Cheney, D.L. and Seyfarth, R.M. (1985). Vervet monkey alarm calls: Manipulation through shared information? *Behaviour*, in press.

Cheney, D.L., Seyfarth, R.M., Andelman, S.J., and Lee, P.C. (1986). Reproductive success in free-ranging vervet monkeys. *In* "Reproductive Success" (T.H. Clutton-Brock, ed.) University of Chicago Press, Chicago, in press.

Chism, J.L. (1978). Relationships between patas infants and group members other than the mother. *In* "Recent Advances in Primatology, Volume 1. Behaviour" (D.J. Chivers and J. Herbert, eds.), pp. 173–176. Academic Press, London.

Chism, J.L., Olson, D.K., and Rowell, T.E. (1982). Reproductive strategies of male patas monkeys. Paper presented at the *Proc. 9th Intern. Primatol. Soc.*, Atlanta, Georgia, August 8–13.

Ciani, A.C. (1984). A case of infanticide in a free-ranging group of rhesus monkeys (*Macaca mulatta*) in the Jackoo Forest, Simla, India. *Primates* **25**, 372–377.

Collins, D.A. (1981). Social behaviour and patterns of mating among adult yellow baboons (*Papio c. cynocephalus*. L. 1766). Ph.D. Thesis, University of Edinburgh.

Collins, D.A. (1985). Interactions between adult male and infant yellow baboons (*Papio c. cynocephalus*) in Tanzania. *Animal Behav.*, in press.

Collins, D.A., Busse, C.D., and Goodall, J. (1984). Infanticide in two populations of savannah baboons. *In* "Infanticide: A Comparative and Evolutionary Perspective" (G. Hausfater and S.B. Hrdy, eds.), pp. 193–215. Aldine, New York.

Cords, M. (1984). Mating patterns and social structure in redtail monkeys (*Cercopithecus ascanius*). *Z. Tierpsychol.* **64**, 313–339.

Cox, K. and LeBoeuf, B.J. (1977). Female incitation of male competition: A mechanism in sexual selection. *Amer. Natur.* **111**, 317–335.

Cronin J.E., and Meikle, D.B. (1982). Hominid and gelada baboon evolution: Agreement between molecular and fossil time scales. *Intern. J. Primatol.* **3**, 469–482.

Cronin, J.E., Cann, R., and Sarich, V.M. (1980). Molecular evolution and systematics of the genus *Macaca*. *In* "The Macaques: Studies in Ecology, Behavior and Evolution" (D.G. Lindburg, ed.), pp. 31–51. Van Nostrand Reinhold, New York.

Curtin, S.H. (1980). Dusky and banded leaf monkeys. *In* "Malayan Forest Primates" (D.J. Chivers, ed.), pp. 107–145. Plenum, New York.

Darwin, C. (1871). "The Descent of Man and Selection in Relation to Sex." John Murray, London.

Dawkins, R. and Krebs, J.R. (1978). Animal signals: Information or manipulation? *In* "Behavioral Ecology: An Evolutionary Approach" (J.R. Krebs and N.B. Davies, eds.), pp. 282–309. Sinauer, Sunderland, Mass.

Deag, J.M. (1977). Aggression and submission in monkey societies. *Animal Behav.* **25**, 465–474.

Defler, T.R. (1982). A comparison of intergroup behavior in *Cebus albifrons* and *C. apella. Primates* **23**, 385–392.

Delson, E. (1980). Fossil macaques, phyletic relationships and a scenario of deployment. *In* "The Macaques: Studies in Ecology, Behavior and Evolution" (D.G. Lindburg, ed.), pp. 10–30. Van Nostrand Reinhold, New York.

DeVore, I. (1962). The social behavior and organization of baboon troops. Ph.D Thesis, University of Chicago.

DeVore, I. (1963). Mother-infant relations in free-ranging baboons. *In* "Maternal Behavior in Mammals" (H. Rheingold, ed.), pp. 305–335. John Wiley, New York.

DeVore, I. (1965). Male dominance and mating behavior in baboons. *In* "Sex and Behavior," (F.A. Beach, ed.), pp. 266–289. John Wiley, New York.

Dittus, W.P. (1979). The evolution of behaviors regulating density and age-specific sex ratios in a primate population. *Behaviour* **69**, 265–297.

Drickamer, L.C. (1974). A ten year summary of reproductive data for free-ranging *Macaca mulatta. Folia Primatol.* **21**, 61–80.

Dunbar, R.I.M. (1980). Determinants and evolutionary consequences of dominance among female gelada baboons. *Behav. Ecol. Sociobiol.* **7**, 253–265.

Dunbar, R.I.M. (1983). Structure of gelada baboon reproductive units. III. The male's relationship with his females. *Animal Behav.* **31**, 565–575.

Dunbar, R.I.M. (1984). Infant-use by male gelada in agonistic contexts: Agonistic buffering, progeny protection, or soliciting support? *Primates* **25**, 28–35.

Dunbar, R.I.M. and Dunbar, E.P. (1975). Social dynamics of gelada baboons. *Contrib. Primatol.* **6**, 1–150. Karger, Basel.

Enomoto, T. (1974). The sexual behavior of Japanese monkeys. *J. Human Evol.* **3**, 351–372.

Enomoto, T. (1981). Male aggression and the sexual behavior of Japanese monkeys. *Primates* **22**, 15–23.

Estrada, A. (1984). Male-infant interactions among free-ranging stumptail macaques. *In* "Primate Paternalism" (D.M. Taub, ed.), pp. 56–87. Van Nostrand Reinhold, New York.

Estrada, A. and Sandoval, J.M. (1977). Social relations in a free-ranging troop of stumptail macaques (*Macaca arctoides*): Male-care behavior I. *Primates* **18**, 793–813.

Fedigan, L.M. (1976). A study of roles in the Arashiyama West troop of Japanese monkeys (*Macaca fuscata*). *Contrib. Primatol.* **9**, 1–95.

Fedigan, L.M. (1982). "Primate Paradigms: Sex Roles and Social Bonds." Eden Press, Montreal.

Fedigan, L.M. and Gouzoules, H. (1978). The consort relationship in a troop of Japanese monkeys. *In* "Recent Advances in Primatology, Volume 1: Behaviour" (D.J. Chivers and J. Herbert, eds.), pp. 493–495. Academic Press, London.

Fisher, R.A. (1930). "The Genetical Theory of Natural Selection." The Clarendon Press, Oxford.
Fossey, D. (1983). "Gorillas in the Mist." Houghton Mifflin, Boston.
Fujii, H. (1975). A psychological study of the social structure of a free-ranging group of Japanese monkeys in Katsuyama. *In* "Contemporary Primatology: Proceedings of the Fifth International Congress of Primatology" (S. Kondo, M. Kawai, and A. Ehara, eds.), pp. 428–436. Karger, Basel.
Gadgil, M. and Bossert, W.H. (1970). Life historical consequences of natural selection. *Amer. Natur.* **104**, 1–24.
Galdikas, B.M.F. and Teleki, G. (1981). Variations in subsistence activities of female and male pongids: New perspectives on the origins of hominid labor division. *Current Anthropol.* **22**, 241–256.
Gaulin, S.J.C. and Sailer, L.D. (1984). Sexual dimorphism in weight among the primates: The relative impact of allometry and sexual selection. *Intern. J. Primatol.* **5**, 515–535.
Gilmore, H. (1980). A syntactic, semantic, and pragmatic analysis of a baboon vocal display. Ph.D. Thesis, University of Pennsylvania.
Goodall, J. (1968). The behaviour of free-living chimpanzees in the Gombe stream area. *Animal Behav. Monogr.* **1**, 161–311.
Goodall, J. (1975). The behaviour of the chimpanzee. *In* "Hominisation und Verhalten" (I. Eibl-Eibesfeld, ed.), pp. 74–136. Gustav Fischer Verlag, Stuttgart.
Goodall, J. (1977). Infant killing and cannibalism in free-living chimpanzees. *Folia Primatol.* **28**, 259–282.
Goodall, J. (1979). Life and death at Gombe. *Natl. Geogr.* **155**, 74–136.
Goodall, J., Bandora, A., Bergmann, E., Busse, C.D., Matama, H., Mpongo, E., Pierce, A., and Riss, D. (1979). Intercommunity interactions in the chimpanzee population of the Gombe National Park. *In* "The Great Apes" (D.A. Hamburg and E.R. McCown, eds.), pp. 13–53. Benjamin/Cummings, Menlo Park, Calif.
Goodall, J. and Hamburg, D.A. (1975). Chimpanzee behavior as a model for the behavior of early man: New evidence on possible origins of human behavior. *In* "American Handbook of Psychiatry: New Psychiatric Frontiers" (D.A. Hamburg and H.K.H. Brodie, eds.), Vol. 6, pp. 14–43. Basic Books, New York.
Gouzoules, H. (1975). Maternal rank and early social interactions of infant stumptail macaques, *Macaca arctoides*. *Primates* **16**, 405–418.
Gouzoules, H. (1980). A description of geneological rank changes in a troop of Japanese monkeys. *Primates* **21**, 262–267.
Gouzoules, H. (1984). Social relations of males and infants in a troop of Japanese macaques: A consideration of causal mechanisms. *In* "Primate Paternalism" (D.M. Taub, ed.), pp. 127–145. Van Nostrand Reinhold, New York.
Gouzoules, H., Gouzoules, S., and Fedigan, L.M. (1982). Behavioural dominance and reproductive success in female Japanese monkeys (*Macaca fuscata*). *Animal Behav.* **30**, 335–349.
Griffin, D.R. (1984). "Animal Thinking." Harvard University Press, Cambridge.
Hall, K.R.L. (1962). The sexual, agonistic and derived social behavior patterns of the wild chacma baboon, *Papio ursinus*. *Proc. Zool. Soc. London* **13**, 283–327.

Hall, K.R.L. (1963). Variations in the ecology of the chacma baboon, *Papio ursinus*. *Symp. Zool. Soc. London* **10**, 1–28.

Hall, K.R.L. (1967). Social interactions of the adult male and adult females of a patas monkey group. *In* "Social Communication among Primates" (S.A. Altmann, ed.), pp. 261–280. University of Chicago Press, Chicago.

Hall, K.R.L. and DeVore, I. (1965). Baboon social behavior. *In* "Primate Behavior: Field Studies of Monkeys and Apes" (I. DeVore, ed.), pp. 53–110. Holt, Rinehart and Winston, New York.

Hamburg, D.A. (1968). Emotions in the perspective of human evolution. *In* "Perspectives on Human Evolution" (S.L. Washburn and P. Jay, eds.), Vol. 1, pp. 246–257. Holt, Rinehart and Winston, New York.

Hamilton, W.J. III. (1984). Significance of paternal investment by primates to the evolution of male–female associations. *In* "Primate Paternalism" (D.M. Taub, ed.), pp. 309–335. Van Nostrand Reinhold, New York.

Hamilton, W.J. III. and Arrowood, P.C. (1978). Copulatory vocalizations of chacma baboons (*Papio ursinus*), gibbons (*Hylobates hoolock*), and humans. *Science* **200**, 1405–1409.

Harcourt, A. (1979). Social relationships between adult male and female mountain gorillas in the wild. *Animal Behav.* **27**, 325–342.

Harding, R.S.O. (1973a). Predation by a troop of olive baboons (*Papio anubis*) *Amer. J. Phys. Anthropol.* **38**, 587–592.

Harding, R.S.O. (1973b). Range utilization by a troop of olive baboons (*Papio anubis*). Ph.D Thesis, University of California, Berkeley.

Harding, R.S.O. (1976). Ranging patterns of a troop of baboons (*Papio anubis*) in Kenya. *Folia Primatol.* **25**, 143–185.

Harding, R.S.O. (1980). Agonism, ranking, and the social behavior of adult male baboons. *Amer. J. Phys. Anthropol.* **53**, 203–216.

Hausfater, G. (1975). Dominance and reproduction in baboons. *Contrib. Primatol.* **7**, 1–150.

Hausfater, G., Altmann, J., and Altmann, S. (1982). Long-term consistency of dominance relations among female baboons (*Papio cynocephalus*). *Science* **217**, 752–755.

Hendrickx, A.G. and Kraemer, D.G. (1969). Observations on the menstrual cycle, optimal mating time and pre-implantation embryos of the baboon, *Papio anubis* and *Papio cynocephalus*. *J. Reprod. Fertil. Suppl.* **6**, 119–128.

Hinde, R.A. (1976). Interactions, relationships and social structure. *Man* **11**, 1–17.

Hinde, R.A. (1977). On assessing the basis of partner preferences. *Behaviour* 62, 1–9.

Hinde, R.A. and Atkinson, S. (1970). Assessing the roles of social partners in maintaining mutual proximity, as exemplified by mother-infant relations in rhesus monkeys. *Animal Behav.* **18**, 169–176.

Hinde, R.A. and Proctor, L.P. (1977). Changes in the relationships of captive rhesus monkeys on giving birth. *Behaviour* **61**, 304–321.

Hiraiwa, M. (1981). Maternal and alloparental care in a troop of free-ranging Japanese monkeys. *Primates* **22**, 309–329.

Hrdy, S.B. (1974). Male–male competition and infanticide among the langurs (*Presbytis entellus*) of Abu, Rajasthan. *Folia Primatol.* **22**, 19–58.

Hrdy, S.B. (1976). Care and exploitation of nonhuman primate infants by conspecifics other than the mother. *In* "Advances in the Study of Behavior"

(J.S. Rosenblatt, R.A. Hinde, E. Shaw, and C. Beer, eds.), Vol. 6, pp. 101–158. Academic Press, New York.

Hrdy, S.B. (1977). "The Langurs of Abu: Female and Male Strategies of Reproduction." Harvard University Press, Cambridge.

Hrdy, S.B. (1979). Infanticide among animals: A review, classification and examination of the implications for the reproductive strategies of individuals. *Ethol. Sociobiol.* **1**, 13–40.

Hrdy, S.B. and Hausfater, G. (1984). Comparative and evolutionary perspectives on infanticide: Introduction and overview. *In* "Infanticide: Comparative and Evolutionary Perspectives" (G. Hausfater and S.B. Hrdy, eds.), pp. xiii–xxxv. Aldine, New York.

Hrdy, S.B. and Williams, G.C. (1983). Behavioral biology and the double standard. *In* "Social Behavior of Female Vertebrates" (S.K. Wasser, ed.), pp. 3–17. Academic Press, New York.

Hunter, J.L. (1978). Guardian behavior in the sooty mangabey, *Cercocebus atys. Amer. J. Phys. Anthropol.* **48**, 407 (abstr.).

Itani, J. (1959). Paternal care in the wild Japanese monkey. *Macaca fuscata. Primates* **2**, 61–93.

Izawa, K. (1980). Social behavior of the wild black-capped capuchin (*Cebus apella*). *Primates* **21**, 443–467.

Johnson, J.A. (1984). Social relationships of juvenile olive baboons. Ph.D. Thesis, University of Edinburgh.

Jolly, C.J. (1966). Introduction to the Cercopithecoidea, with notes on their use as laboratory animals. *Symp. Zool. Soc. London* **17**, 427–457.

Jolly, C.J. and Brett, F.L. (1973). Genetic markers and baboon biology. *J. Med. Primatol.* **2**, 85–99.

Judge, P.G. (1982). Redirection of aggression based on kinship in a captive group of pigtail macaques. *Intern. J. Primatol.* **3**, 301 (abstr.).

Kano, T. (1982). The social group of pygmy chimpanzees (*Pan paniscus*) of Wamba. *Primates* **23**, 171–188.

Kaplan, J.R. (1977). Patterns of fight interference in free-ranging rhesus monkeys. *Amer. J. Phys. Anthropol.* **17**, 279–288.

Kaplan, J.R. (1978). Fight interference and altruism in rhesus monkeys. *Amer. J. Phys. Anthropol.* **49**, 241–250.

Kaufman, J.H. (1965). A three-year study of mating behavior of a free-ranging band of rhesus monkeys. *Ecology* **40,** 500–512.

Kaufman, J.H. (1967). Social relations of adult males in a free-ranging band of rhesus monkeys. *In* "Social Communication among Primates" (S.A. Altmann, ed.), pp. 73–98. University of Chicago Press, Chicago.

Kawai, M. (1958). On the system of social ranks in a natural group of Japanese monkeys: Basic rank and dependent rank. *Primates* 1, 111–148 (in Japan.).

Kawai, M. (1965). On the system of social ranks in a natural group of Japanese monkeys: Basic rank and dependent rank. *In* "Japanese Monkeys: A Collection of Translations" (K. Imanishi and S.A. Altmann, eds.), pp. 66–86. Published by the editors, Chicago.

Kawamura, S. (1958). Matriarchal social ranks in the Minoo-B troop: A study of the rank system of Japanese monkeys. *Primates* **1**, 149–156 (in Japan.).

Kawamura, S. (1965). Matriarchal social ranks in the Minoo-B troop: A study of the rank system of Japanese monkeys. *In* "Japanese Monkeys: A collection of Translations" (K. Imanishi and S.A. Altmann, eds.), pp. 105–112. Published by the editors, Chicago.

Kitamura, K. (1983). Pygmy chimpanzee association patterns in ranging. *Primates* **24**, 1–12.

Kleiman, D. (1977). Monogamy in mammals. *Quart. Rev. Biol.* **52**, 36–69.

Koyama, N. (1970). Changes in dominance rank and division of a wild Japanese monkey troop in Arashiyama. *Primates* **11**, 335–390.

Krebs, J.R. and Davies, N.B. (1981). "An Introduction to Behavioral Ecology." Sinauer, Sunderland, Mass.

Kummer, H. (1968). "Social Organization in Hamadryas Baboons." University of Chicago Press, Chicago.

Kummer, H. (1975). Rules of dyad and group formation among captive gelada baboons (*Theropithecus gelada*). *In* "Proceedings from the Symposia of the Fifth Congress of the International Primatological Society" (S. Kondo, M. Kawai, A. Ehara, and S. Kawamura, eds.), pp. 129–159. Japan Science Press, Tokyo.

Kummer, H. (1978). On the value of social relationships to nonhuman primates: A heuristic scheme. *Soc. Sci. Inform.* **17**, 687–705.

Kummer, H. (1982). Social knowledge in free-ranging primates. *In* "Animal Mind–Human Mind" (D.R. Griffin, ed.), pp. 113–130. Springer-Verlag, Berlin.

Kummer, H. (1984). From laboratory to desert and back: A social system of hamadryas baboons. *Animal Behav.* **32**, 965–971.

Kummer, H., Abegglen, J.J., Bachmann, C., Falett, J., and Sigg, H. (1978). Grooming relationships and object competition among hamadryas baboons. *In* "Recent Advances in Primatology, Volume 1: Behaviour" (D.J. Chivers and J. Herbert, eds.), pp. 31–38. Academic Press, London.

Kummer, H., Gotz, W., and Angst, W. (1974). Triadic differentiation: An inhibitory process protecting pair bonds in baboons. *Behaviour* **49**, 62–87.

Kurland, J.A. (1977). Kin selection in the Japanese monkey. *Contrib. Primatol.* **12**, 1–145.

Kurland, J.A. and Gaulin, S.J.C. (1984). The evolution of male parental investment: Effects of genetic relatedness and feeding ecology on the allocation of reproductive effort. *In* "Primate Paternalism" (D.M. Taub, ed.), pp. 259–308. Van Nostrand Reinhold, New York.

Lancaster, J.B. and Lancaster, C.S. (1983). Parental investment: The hominid adaptation. *In* "How Humans Adapt: A Biocultural Odyssey" (D.J. Ortner, ed.), pp. 33–56. Smithsonian, Washington, D.C.

Leighton, D. (1986). Gibbons: Territoriality and monogamy. *In* "Primate Societies" (B. Smuts, D. Cheney, R. Seyfarth, R. Wrangham, and T. Struhsaker, eds.). University of Chicago Press, Chicago, in press.

Leland, L., Struhsaker, T.T., and Butynski, T.M. (1984). Infanticide by adult males in three primate species of Kibale Forest, Uganda: A test of hypotheses. *In* "Infanticide: Comparative and Evolutionary Perspectives" (G. Hausfater and S.B. Hrdy, eds.), pp. 151–172. Aldine, New York.

Lindburg, D.G. (1971). The rhesus monkey in North India: An ecological and behavioral study. *In* "Primate Behavior: Developments in Field and Laboratory Research" (L.A. Rosenblum, ed.), Vol. 2, pp. 2–106. Academic Press, New York.

Lindburg, D.G. (1973). Grooming as a regulator of social interaction in rhesus macaques. *In* "Behavioral Regulators of Behavior in Primates" (C.R.

Carpenter, ed.), pp. 124–148. Bucknell University Press, Lewisburg, Penn.

Lindburg, D.G. (1983). Mating behavior and estrus in the Indian rhesus monkey. *In* "Perspectives in Primate Biology" (P.K. Seth, ed.), pp. 45–61. Today and Tomorrow's Printers and Publishers, New Delhi.

Lovejoy, C.O. (1981). The origin of man. *Science* **211**, 341–350.

McGrew, W.C. 1981a. The female chimpanzee as a human evolutionary prototype. *In* "Woman the Gatherer" (F. Dahlberg, ed.), pp. 35–73. Yale University Press, New Haven.

McGrew, W.C. 1981b. Social and cognitive capabilities of nonhuman primates: Lessons from the wild to captivity. *Intern. J. Study Animal Problems* **2**, 138–149.

MacLennan, A.M. and Wynn, R.M. (1971). Menstrual cycle of the baboon. I. Clinical features, vaginal cytology and endometrial histology. *Obstet. Gynecol.* **38**, 350–358.

Manzolillo, D.L. (1982). Intertroop transfer by adult male *Papio anubis*. Ph.D. Thesis, University of California, Los Angeles.

Massey, A. (1977). Agonistic aids and kinship in a group of pigtail macaques. *Behav. Ecol. Sociobiol.* **2**, 31–40.

Maynard Smith, J. and Parker, G.A. (1976). The logic of asymmetric contests. *Animal Behav.* **24**, 159–175.

Melnick, D. and Pearl, M. (1986). Cercopithecines in multi-male groups: Genetic diversity and population structure. *In* "Primate Societies" (B. Smuts, D. Cheney, R. Seyfarth, R. Wrangham, and T. Struhsaker, eds.). University of Chicago Press, Chicago, in press.

Menzel, E. (1979). Communication of object-locations in a group of young chimpanzees. *In* "The Great Apes" (D.A. Hamburg and E.R. McCown, eds.), pp. 357–371. Benjamin/Cummings, Menlo Park, Calif.

Michael, R.P., Bonsall, R.W., and Zumpe, D. (1978). Consort bonding and operant behavior by female rhesus monkeys. *J. Comp. Physiol. Psychol.* **92**, 837–845.

Michael, R.P., Herbert, J. and Welegalla, J. (1966). Ovarian hormones and grooming behaviour in the rhesus monkey (*Macaca mulatta*) under laboratory conditions. *J. Endocrinol.* **36**, 263–279.

Mitani, J.C. (1985). Mating behaviour of male orangutans in the Kutai Game Reserve, Indonesia. *Animal Behav.*, **33**, 392–402.

Mitchell, C. and Brandt, E.M. (1972). Paternal behavior in primates. *In* "Primate Socialization" (F. Poirier, ed.), pp. 173–206. Random House, New York.

Mori, U. (1979a). Development of sociability and social status. *Contrib. Primatol.* **16**, 125–154.

Mori, U. (1979b). Individual relationships within a unit. *Contrib. Primatol.* **16**, 93–124.

Mori, U. (1979c). Unit formation and the emergence of a new leader. *Contrib. Primatol.* **16**, 155–181.

Nagel, U. (1973). A comparison of anubis baboons, hamadryas baboons and their hybrids at a species border in Ethiopia. *Folia Primatol.* **19**, 104–165.

Napier, J.R. and Napier, P.H. (1967). "A Handbook of Living Primates." Academic Press, London.

Nicolson, N. (1982). Weaning and the development of independence in olive baboons. Ph.D. Thesis, Harvard University.

Nishida, T. (1979). The social structure of chimpanzees of the Mahale Mountains. *In* "The Great Apes" (D.A. Hamburg and E.R. McCown, eds.), pp. 73–121. Benjamin/Cummings, Menlo Park, Calif.

Nishida, T. (1983). Alloparental behavior in wild chimpanzees of the Mahale Mountains, Tanzania. *Folia Primatol.* **41**, 1–33.

Nishida, T., Hiraiwa-Hasegewa, M., Hasegewa, T., and Takahata, Y. (1985). Group extinction and female transfer in wild chimpanzees in the Mahale Mountains. *Z. Tierpsychol.*, **67**, 284–301.

Oki, J. and Maeda, J. (1973). Grooming as a regulator of behavior in Japanese macaques. *In* "Behavioral Regulators of Behavior in Primates" (C.R. Carpenter, ed.), pp. 149–163. Bucknell University Press, Lewisburg, Penn.

Packer, C. (1977). Reciprocal altruism in *Papio anubis. Nature (London)* **265**, 441–443.

Packer, C. (1979a). Inter-troop transfer and inbreeding avoidance in *Papio anubis. Animal Behav.* **27**, 1–36.

Packer, C. (1979b). Male dominance and reproductive activity in *Papio anubis. Animal Behav.* **27**, 37–46.

Packer, C. (1980). Male care and exploitation of infants in *Papio anubis. Animal Behav.* **28**, 521–527.

Packer, C. and Pusey, A.E. (1979). Female aggression and male membership in troops of Japanese macaques and olive baboons. *Folia Primatol.* **31**, 212–218.

Packer, C. and Pusey, A.E. (1985). Asymmetric contests in social mammals: Respect, manipulation and age-specific aspects. *In* "Evolution: Essays in Honour of John Maynard Smith" (P.J. Greenwood, P.H. Harvey, and M. Slatkin, eds.), pp. 173–186. Cambridge University Press, Cambridge.

Pereira, M. (1983). Abortion following the immigration of an adult male baboon (*Papio cynocephalus*). *Amer. J. Primatol.* **4**, 93–98.

Popp. J.L. (1978). Male baboons and evolutionary principles. Ph.D. Thesis, Harvard University.

Popp, J.L. and DeVore, I. (1979). Aggressive competition and social dominance theory: Synopsis. *In* "The Great Apes" (D.A. Hamburg and E.R. McCown, eds.), pp. 317–338. Benjamin/Cummings, Menlo Park, Calif.

Pusey, A.E. (1980). Inbreeding avoidance in chimpanzees. *Animal Behav.* **28**, 543–552.

Ransom, T.W. (1971). Ecology and social behavior of baboons (*Papio anubis*) at the Gombe National Park. Ph.D Thesis, University of California, Berkeley.

Ransom, T.W. (1981). "Beach Troop of the Gombe." Bucknell University Press, Lewisburg, Penn.

Ransom, T.W. and Ransom, B.S. (1971). Adult male-infant relations among baboons (*Papio anubis*). *Folia Primatol.* **16**, 179–195.

Ransom, T.W. and Rowell, T.E. (1972). Early social development in feral baboons. *In* "Primate Socialization" (F.A. Poirier, ed.), pp. 105–144. Random House, New York.

Rasmussen, K.L.R. (1980). Consort behavior and mate selection in yellow baboons (*Papio cynocephalus*). Ph.D. Thesis, University of Cambridge.

Rasmussen, K.L.R. (1983a). Age-related variation in the interactions of adult females with adult males in yellow baboons. *In* "Primate Social Relationships: An Integrated Approach" (R.A. Hinde, ed.), pp. 47–53. Blackwell, Oxford.

Rasmussen, K.L.R. (1983b). Influence of affiliative preferences upon the behaviour of male and female baboons during sexual consortships. *In* "Primate Social Relationships: An Integrated Approach" (R.A. Hinde, ed.), pp. 116–120. Blackwell, Oxford.

Rawlins, R.G., Kessler, M.J., and Turnquist, J.E. (1984). Reproductive performance, population dynamics and anthropometrics of the free-ranging Cayo Santiago rhesus macaques. *J. Med. Primatol.* **13**, 247–259.

Redican, W.K. (1976). Adult male–infant interactions in nonhuman primates. *In* "The Role of the Father in Child Development" (M.E. Lamb, ed.), pp. 345–385. John Wiley, New York.

Reynolds, P.C. (1981). "On the Evolution of Human Behavior." University of California Press, Berkeley.

Robinson, J.G. (1981). Spatial structure in foraging groups of wedge-capped capuchin monkeys *Cebus nigrivittatus. Animal Behav.* **29**, 1036–1056.

Robinson, J.G., Wright, P.C., and Kinzey, W.G. (1986). Monogamous cebids and their relatives: Intergroup calls and spacing. *In* "Primate Societies" (B. Smuts, D. Cheney, R. Seyfarth, R. Wrangham, and T. Struhsaker, eds.). University of Chicago Press, Chicago, in press.

Rowell, T.E. (1967). A quantitative comparison of the behaviour of a wild and a caged baboon group. *Animal Behav.* **15**, 499–509.

Rowell, T.E., (1974). Contrasting adult male roles in different species of nonhuman primates. *Arch. Sexual Behav.* **3**, 143–149.

Rowell, T.E., Din, N.A. and Omar, A. (1968). The social development of baboons in their first three months. *J. Zool. (London)* **155**, 461–483.

Saayman, G.S. (1970). The menstrual cycle and sexual behavior in a troop of free-living chacma baboons *(Papio ursinus). Folia Primatol.* **12**, 81–110.

Saayman, G.S. (1971a). Behaviour of the adult males in a troop of free-ranging chacma baboons *(Papio ursinus). Folia Primatol.* **15**, 36–57.

Saayman, G.S. (1971b). Grooming behavior in a troop of chacma baboons. *Folia Primatol.* **16**, 161–178.

Sade, D.S. (1965). Some aspects of parent–offspring and sibling relations in a group of rhesus monkeys, with a discussion of grooming. *Amer. J. Phys. Anthropol.* **23**, 1–17.

Sade, D.S. (1967). Determinants of dominance in a group of free-ranging rhesus monkeys. *In* "Social Communication among Primates" (S.A. Altmann, ed.), pp. 99–114. University of Chicago Press, Chicago.

Sade, D.S., Cushing, K., Cushing, P., Dunaif, J., Figueroa, A., Kaplan, J., Lauer, C., Rhodes, D., and Schneider, J. (1976). Population dynamics in relation to social structure on Cayo Santiago. *Yearb. Phys. Anthropol.* **20**, 253–262.

Sapolsky, R.M. (1983). Endocrine aspects of social instability in the olive baboon *(Papio anubis). Amer. J. Primatol.* **5**, 365–379.

Scott, L.M. (1984). Reproductive behavior of adolescent female baboons (*Papio anubis*) in Kenya. *In* "Female Primates: Studies by Women Primatologists" (M. Small, ed.), pp. 77–100. Alan R. Liss, New York.

Seyfarth, R.M. (1976). Social relationships among adult female baboons. *Animal Behav.* **24**, 917–938.

Seyfarth, R.M. (1977). A model of social grooming among adult female monkeys. *J. Theor. Biol.* **65**, 671–698.

Seyfarth, R.M. (1978a). Social relationships among adult male and female baboons, I. Behaviour during sexual consortship. *Behaviour* **64**, 204–226.

Seyfarth, R.M. (1978b). Social relationships among adult male and female baboons, II. Behaviour throughout the female reproductive cycle. *Behaviour* **64**, 227–247.

Seyfarth, R.M. (1980). The distribution of grooming and related behaviours among adult female vervet monkeys. *Animal Behav.* **28**, 798–813.

Seyfarth, R.M. and Cheney, D.L. (1984). Grooming, alliances and reciprocal altruism in vervet monkeys. *Nature (London)* **308**, 541–543.

Shaikh, A., Celaya, C., Gomez, I., and Shaikh, S. (1982). Temporal relationship of hormonal peaks to ovulation and sex skin deturgescence in the baboon. *Primates* **22**, 444–452.

Shopland, J.M. (1982). An intergroup encounter with fatal consequences in yellow baboons (*Papio cynocephalus*). *Amer. J. Primatol.* **3**, 263–266.

Shotake, T. (1981). Population genetical study of natural hybridization between *Papio anubis* and *Papio hamadryas*. *Primates* **22**, 285–308.

Sigg, H. (1980). Differentiation of female positions in hamadryas one-male units. *Z. Tierpsychol.* **53**, 265–302.

Sigg, H., Stolba, A., Abegglen, J.J., and Dasser V. (1982). Life history of hamadryas baboons: Physical development, infant mortality, reproductive parameters and family relationships. *Primates* **23**, 473–487.

Silk, J.B. (1980). Kidnapping and female competition among captive bonnet macaques. *Primates* **21**, 100–110.

Silk, J.B., Clark-Wheatley, C.B., Rodman, P.S., and Samuels, A. (1981a). Differential reproductive success and facultative adjustment of sex ratios among female bonnet macaques *(Macaca radiata)*. *Animal Behav.* **29**, 1106–1120.

Silk, J.B., Rodman, P.S., and Samuels, A. (1981b). Hierarchical organization of female *Macaca radiata*. *Primates* **22**, 84–95.

Silk, J.B., Samuels, A., and Rodman, P.S. (1981c). The influence of kinship, rank, and sex on affiliation and aggression between adult females and immature bonnet macaques (*Macaca radiata*). *Behaviour* **78**, 111–137.

Smith, E.O. and Peffer-Smith, P.G. (1984). Adult male–immature interactions in captive stumptail macaques (*Macaca arctoides*). *In* "Primate Paternalism" (D.M.Taub, ed.), pp. 88–112. Van Nostrand Reinhold, New York.

Smuts, B.B. (1980). Effects on social behavior of loss of high rank in wild adult female baboons (*Papio anubis*). Paper presented at the *Annu. Meeting Animal Behav. Soc.*, Fort Collins, Colorado, June 9–13.

Smuts, B.B. (1982). Special relationships between adult male and female olive baboons (*Papio anubis*). Ph.D. Thesis, Stanford University.

Smuts, B.B. (1983a). Dynamics of special relationships between adult male and female olive baboons. *In* "Primate Social Relationships: An Integrated Approach" (R.A. Hinde, ed.), pp. 112–116. Blackwell, Oxford.

Smuts, B.B. (1983b). Special relationships between adult male and female olive baboons: Selective advantages. *In* "Primate Social Relationships: An Integrated Approach" (R.A. Hinde, ed.), pp. 262–266. Blackwell, Oxford.

Smuts, B.B. (1985). "Social Relationships among Adult Female Olive Baboons." In preparation.

Smuts, B.B. and Watanabe, J.M. (1985). "Social Relationships among Adult Male Olive Baboons: Greetings and Coalitions." In preparation.

Snowdon, C.T., and Suomi, S.J. (1982). Paternal behavior in primates. *In* "Child Nurturance, Volume 3. Studies of Development in Nonhuman

Primates" (H. Fitzgerald, J. Mullins, and P. Gage, eds.), pp. 63–108. Plenum, New York.

Stein, D.M. (1981). The nature and function of social interactions between infant and adult male yellow baboons (*Papio cynocephalus*). Ph.D. Thesis, University of Chicago.

Stein, D.M. (1984). Ontogeny of infant–adult male relationships during the first year of life for yellow baboons (*Papio cynocephalus*). *In* "Primate Paternalism" (D.M. Taub, ed.), pp. 213–243. Van Nostrand Reinhold, New York.

Stein, D.M. and Stacey, P.B. (1981). A comparison of infant–adult male relations in a one-male group with those in a multi-male group for yellow baboons (*Papio cynocephalus*). *Folia Primatol.* **36**, 264–276.

Stevenson-Hinde, J. (1983). Individual characteristics: A statement of the problem. *In* "Primate Social Relationships: An Integrated Approach" (R.A. Hinde, ed.), pp. 28–30. Blackwell, Oxford.

Strassman, B.I. (1981). Sexual selection, paternal care, and concealed ovulation in humans. *Ethol. Sociobiol.* **2**, 31–40.

Struhsaker, T.T. (1967). Social structure among vervet monkeys (*Cercopithecus aethiops*). *Behaviour* **29**, 83–121.

Strum, S.C. (1975). Life with the Pumphouse Gang: New insights into baboon behavior. *Nat. Geogr.* **147**, 672–691.

Strum, S.C. (1981). Processes and products of change: Baboon predatory behavior at Gilgil, Kenya. *In* "Omnivorous Primates: Gathering and Hunting in Human Evolution" (R.S.O. Harding and G. Teleki, eds.), pp. 255–302. Columbia University Press, New York.

Strum, S.C. (1982). Agonistic dominance in male baboons: An alternative view. *Intern. J. Primatol.* **3**, 175–202.

Strum, S.C. (1983). Use of females by male olive baboons (*Papio anubis*). *Amer. J. Primatol.* **5**, 93–109.

Strum, S.C. (1984). Why males use infants. *In* "Primate Paternalism" (D.M. Taub, ed.), pp. 146–185. Van Nostrand Reinhold, New York.

Strum, S.C. and Western, D. (1982). Variations in fecundity with age and environment in olive baboons (*Papio anubis*). *Amer. J. Primatol.* **3**, 61–76.

Sugiyama, Y. (1965). On the social change of hanuman langurs (*Presbytis entellus*) in their natural condition. *Primates* **6**, 381–418.

Sugiyama, Y. (1971). Characteristics of the social life of bonnet macaques (*Macaca radiata*). *Primates* **12**, 247–266.

Sugiyama, Y. and Ohsawa, H. (1982). Population dynamics of Japanese monkeys with special reference to the effect of artificial feeding. *Folia Primatol.* **39**, 238–263.

Symons, D. (1979). "The Evolution of Human Sexuality." Oxford University Press, New York.

Takahata, Y. (1982a). Social relations between adult males and females of Japanese monkeys in the Arashiyama B Troop. *Primates* **23**, 1–23.

Takahata, Y. (1982b). The socio-sexual behavior of Japanese monkeys. *Z. Tierpsychol.* **59**, 89–108.

Tanner, N.M. (1981). "On Becoming Human." Cambridge University Press, Cambridge.

Tanner, N.M. and Zihlman, A.L. (1976). Women in evolution. Part 1. Innovation and selection in human origins. *Signs* **1**, 585–608.

Taub, D.M. (1980). Female choice and mating strategies among wild Barbary macaques (*Macaca sylvanus*). *In* "The Macaques: Studies in Ecology, Behavior and Evolution" (D.G. Lindburg, ed.), pp. 287–344. Van Nostrand Reinhold, New York.

Taub, D.M. (1984). Male caretaking behavior among wild Barbary macaques (*Macaca sylvanus*). *In* "Primate Paternalism" (D.M. Taub, ed.), pp. 20–55. Van Nostrand Reinhold, New York.

Taylor, H., Teas, J., Richie, T., Southwick, C., and Shrestha, R. (1978). Social interactions between adult male and infant rhesus monkeys in Nepal. *Primates* **19**, 343–351.

Tilford, B.L. and Nadler, R.D. (1978). Male parental behavior in a captive group of lowland gorillas (*Gorilla gorilla gorilla*). *Folia Primatol.* **29**, 218–228.

Tokuda, K. (1961). A study on the sexual behavior in the Japanese monkey troop. *Primates* **3**, 2–40.

Trivers, R.L. (1971). The evolution of reciprocal altruism. *Quart. Rev. Biol.* **46**, 35–57.

Trivers, R.L. (1972). Parental investment and sexual selection. *In* "Sexual Selection and the Descent of Man, 1871–1971" (B. Campbell, ed.), pp. 136–179. Aldine, Chicago.

Trivers, R.L. (1985). "Social Evolution." Benjamin/Cummings, Menlo Park, Calif.

Tsingalia, H.M. and Rowell, T.E. (1984). The behaviour of adult male blue monkeys. *Z. Tierpsychol.* **64**, 253–268.

Turke, P.W. (1984). Effects of ovulatory concealment and synchrony on protohominid mating systems and parental roles. *Ethol. Sociobiol.* **5**, 33–44.

Tutin, C.E.G. (1979). Mating patterns and reproductive strategies in a community of wild chimpanzees (*Pan troglodytes schweinfurthii*). *Behav. Ecol. Sociobiol.* **6**, 29–38.

Tutin, C.E.G. (1980). Reproductive behaviour of wild chimpanzees in the Gombe National Park, Tanzania. *J. Reprod. Fertil. Supp.* **28**, 43–57.

Vessey, S.H. and Meikle, D.B. (1984). Free-living rhesus monkeys: Adult male interactions with infants and juveniles. *In* "Primate Paternalism" (D.M. Taub, ed.), pp. 113–126. Van Nostrand Reinhold, New York.

Vogt, J.L. (1984). Interactions between adult males and infants in prosimians and New World monkeys. *In* "Primate Paternalism" (D.M. Taub, ed.), pp. 346–376. Van Nostrand Reinhold, New York.

Waal, F.B.M. de. (1977). The organization of agonistic relations within two captive groups of Java-monkeys (*Macaca fascicularis*). *Z. Tierpsychol.* **44**, 225–282.

Waal, F.B.M. de. (1982). "Chimpanzee Politics: Power and Sex among Apes." Harper and Row, New York.

Waal, F.B.M. de. (1986). Dynamics of social relationships. *In* "Primate Societies" (B. Smuts, D. Cheney, R. Seyfarth, R. Wrangham, and T. Struhsaker, eds.), University of Chicago Press, Chicago, in press.

Walters, J. (1980). Interventions and the development of dominance relationships in female baboons. *Folia Primatol.* **34**, 61–89.

Washburn, S.L. and DeVore, I. (1963). The social life of baboons. *In* "Primate Social Behavior" (C.H. Southwick, ed.), pp. 98–113. Van Nostrand Reinhold, New York.

Wasser, S.K. (1983). Reproductive competition and cooperation among female yellow baboons. *In* "Social Behavior of Female Vertebrates" (S.K. Wasser, ed.), pp. 350–390. Academic Press, New York.

Wasser, S.K. and Barash, D.P. (1983). Reproductive suppression among female mammals: Implications for biomedicine and sexual selection theory. *Quart. Rev. Biol.* **58**, 513–538.

Watanabe, K. (1979). Alliance formation in a free-ranging troop of Japanese macaques. *Primates* **20**, 459–474.

Whitten, P.L. (1983). Diet and dominance among female vervet monkeys (*Cercopithecus aethiops*). *Amer. J. Primatol.* **5**, 139–159.

Whitten, P.L. (1986). Infants and adult males. *In* "Primate Societies" (B. Smuts, D. Cheney, R. Seyfarth, R. Wrangham, and T. Struhsaker, eds.). University of Chicago Press, Chicago, in press.

Wildt, D.E., Doyle, L.L., Stone, S.C., and Harrison, R.M. (1977). Correlation of perineal swelling with serum ovarian hormone levels, vaginal cytology, and ovarian follicular development during the baboon reproductive cycle. *Primates* **18**, 261–270.

Williams, G.C. (1966). "Adaptation and Natural Selection." Princeton University Press, Princeton.

Wilson, M.J. (1982). Behavior as a determinant of kinship in olive baboons (*Papio anubis*). Undergraduate Honors Thesis, Department of Anthropology, Harvard University.

Woodruff, G. and Premack, D. (1979). Intentional communication in the chimpanzee: The development of deception. *Cognition* **7**, 333–362.

Wrangham, R.W. (1975). Behavioural ecology of chimpanzees in Gombe National Park, Tanzania. Ph.D. Thesis, University of Cambridge.

Wrangham, R.W. (1979a). On the evolution of ape social systems. *Social Sci. Inform.* **18**, 335–368.

Wrangham, R.W. (1979b). Sex differences in chimpanzee dispersion. *In* "The Great Apes" (D.A. Hamburg and E.R. McCown, eds.), pp. 481–489. Benjamin/Cummings, Menlo Park, Calif.

Wrangham, R.W. (1981). Drinking competition in vervet monkeys. *Animal Behav.* **29**, 904–910.

Wrangham, R.W. (1986). Ecology and social relationships in two species of chimpanzee. *In* "Social Ecology in Birds and Mammals" (D.I. Rubenstein and R.W. Wrangham, eds.). Princeton University Press, Princeton, in press.

Wrangham, R.W. and Smuts, B.B. (1980). Sex differences in the behavioural ecology of chimpanzees in the Gombe National Park, Tanzania. *J. Reprod. Fertil. Suppl.* **28**, 43–65.

Zihlman, A.L. (1978). Women in evolution. Part 2. Subsistence and social organization among early hominids. *Signs* **4**, 4–20.

Zihlman, A.L. (1981). Women as shapers of the human adaptation. *In* "Woman the Gatherer" (F. Dahlberg, ed.), pp. 75–120. Yale University Press, New Haven.

SUBJECT INDEX

A

G

H

I

N

P

R

S